TESTING ELECTRICAL INS
A PRACTICAL GUIDE FOR
SEVENTH EDI

By Anthony Hinsley MA

COPYRIGHTS, ACKNOWLEDGEMENTS

First, second, fourth, fifth and sixth Edition published in England from 1996 - 2005
Seventh edition published in France 2006.

Every effort has been made to ensure the information and guidance in this book is correct and accurate at the time of going to press. However, it is understood that all parties must use their skill, knowledge and understanding when making use of it.

Neither the author or the publisher hold themselves responsible for any consequences that may arise from errors or omissions and negate any liabilities.

The drawings of instruments in this book are graphical representations of concept meters and are used so as not to reflect any particular manufacturer, make or model, allowances must be made for their realism.

I would like to thank my wife Sally and P & R Hurt Education & Training for their input.

ISBN 2-9524138-3-5

CONTENTS

TESTING ELECTRICAL INSTALLATIONS
A PRACTICAL GUIDE FOR ELECTRICIANS

Introduction

To introduce myself, I am a qualified Electrician with 30 years' experience in the industry, now creating and supplying educational and training resources to the Electrical Installation industry for people who want to be Electricians or make a career in the Electrical industry, and I address this book to training and practising Electricians.

Throughout the text I will address all my readers as "You" just as I would if we were in verbal discussion together as I believe that this will make for easier reading and understanding .
"Ok, you are a practising Electrician, but can you test an electrical installation thoroughly and correctly in accordance with the requirements of BS 7671"?

For instance, during my years in education I found that a majority of Electricians can indeed carry out an insulation resistance test or a continuity test, but very few practising Electricians could test most installations thoroughly and correctly in accordance with the requirements of BS 7671.

Do you know,

What the tests are for?

Why you are carrying out the tests?

What do the results mean?

How to fill out test reports and certificates correctly.

If your answer to any of these questions is no, then I hope that after reading and putting into practice the tests in this book, you will be confident that you have the necessary competence to carry out thorough testing. Of course, as with all skills, one has to practice over a period of time to become proficient.

If your training was initially to National standards then you know that in order to be deemed competent you must carry out each task satisfactorily and consistently before competency can be achieved, and as a practising Electrician you should be aware of the current edition of the Regulations.

Regulation 711-11-01 states "Every installation shall, during erection and/or completion before being put into service be inspected and tested to verify, so far as is reasonably practicable, that the requirements of the Regulations have been met". Remember; You must test the installation as you install it.

TEST EQUIPMENT

Firstly let's deal with the most frequently used test instruments.

Insulation/continuity tester:

The instrument allows the measurement of the resistance in a circuit in Ωs and can generate voltages to detect insulation leakage between live conductors and live conductors to earth.
Continuity testing at open circuit voltages between 4V and 24V a.c. or d.c. and able to deliver a closed circuit current of 200mA for this purpose.
Insulation testing at voltages of 250V, 500V and 1000V dc, at 1mA for 1 minute.

Residual current device tester:

This instrument is capable of measuring the time taken for RCDs to trip under earth fault conditions. It should be capable of testing residual current devices on the following ranges.

1/2X or 50% I Δ N of the rated tripping current of the device.
1 X or 100% I Δ N of the rated tripping current of the device.
150mA current test (see page 78 note 6)

These tests should also be carried out on the 0° and 180° phase angle to test that tripping is as efficient in the positive section of the wave form as it is in the negative section. The facility for testing d.c. operated trips is desirable but not an absolute requirement. You should also have a special lead set to enable you to test circuits other than sockets outlets.

Earth loop impedance tester:

This instrument is designed to measure the resistance of the circuit utilising the voltage and applying fault currents up to 25A. When you carry out the test it internally carries out calculations to display the resistance in ohms. This instrument also allows the calculation or measurement of **PSCC (prospective short circuit current)** Phase to Neutral and **PEFC (prospective earth fault current)** Phase to Earth. You should also have a special lead set to enable you to test circuits other than sockets outlets.

Note; HSE GS38 specifies test equipment for electricians.

HEALTH AND SAFETY

Before you begin testing an installation or individual circuits, there are safety issues that need to be addressed.

In the regulations it states 'Precautions shall be taken to avoid danger to persons and to avoid damage to property and installed equipment during inspection and testing' (711-01-01).

This regulation is very broad and covers just about every conceivable occurrence that could occur. Let's look at the regulation in two parts.

1] 'Precautions shall be taken to avoid danger to persons'.
Think of the following points before carrying out testing:

- 'Have you isolated the circuit correctly?'
- 'Are you using long test leads that someone could trip over?'
- 'Have you informed people of the possibility of shock whilst you are carrying out insulation testing?'
- 'Have you checked to see if any essential services are fed from the board or circuit you intend to test, such as emergency lighting, fire alarms, life-support equipment in hospitals, gas monitoring systems, standby generators, UPS systems, etc.,?'
- 'Have you discharged any capacitors in the circuit?'
- 'Are your instruments of the appropriate type to be testing in any particular environment, i.e. are you using intrinsically safe test equipment in potentially explosive areas?'

2] 'Avoid damage to property and installed equipment during inspection and testing'.

- 'Is there an RCD protecting the circuit?'
- 'Are there computers on line?'
- 'Is there electronic equipment in the circuit?'
  When removing covers to equipment 'are there any special gaskets that may need to be replaced?'
- 'Are all loads disconnected?'
- 'Are there any mixing, heating, ultra violet or other forms of processing that needs consideration for the length of time that it may be disconnected?'
- 'Is air conditioning or extraction essential to the working environment?'

This page gives an example of the sort of questions you should be asking yourself... Remember **SAFETY FIRST!**

SAFE ISOLATION

Electrical supplies above extra low voltage are potentially very dangerous and can kill huma beings and livestock and therefore should be treated with the greatest respect.

It may be possible to carry out some of the initial verification (visual inspection) tests withou disconnecting the supply but for part of the inspection it will be necessary to isolate the supply Under British Standards (BS 6423) there is a laid down procedure for isolating installations o circuits to allow work to be carried out safely, thus preventing electric shock through inadverten re-instatement of the supply.

Refer to group of drawings Fig: 1A, Fig: 1B and Fig: 1C on page 5

In brief, the sequence is as follows:

- Locate circuit/circuits to be isolated
- Remove all loads (unplug and switch off)
- Test circuit to establish mains voltage using an approved voltage indicator (Fig:1A)
- Isolate circuit/circuits and observe voltage indicator (Fig:1B)
- Lock off and/or remove fuses (retain fuses about your person) (Fig:1B)
- Test circuit/circuits are dead. Test between P-N, P-E, N-E for 1 phase and L1-L2, L2-L3, L1-L3, & L1-N, L2-N, L3-N, L1-E, L2-E, L3-E for 3 phase
- Test voltage indicator on known supply and or proving unit (Fig:1C)
- Erect warning notices stating **'DANGER ELECTRICIAN AT WORK'**

Remember that BS 6423 states that all **live conductors must be isolated** before work can be carrie out. As the neutral conductor is classified as a live conductor this should also be disconnecte (Fig:1B) to comply with the British Standard. This may mean removing the conductor from th neutral bar in distribution boards. Not all distribution boards are fitted with double pole isolators s the connecting sequence for neutral conductors needs to be verified and maintained (see page 7 note 3).

You should consider the use of a paper based permit to work system for all isolation procedures Also note on the warning notice **'DANGER ELECTRICIAN AT WORK'** where you are working your name, the time and date you started work. This will help anyone to establish why power ha been turned off.

Testing for the presence of voltage

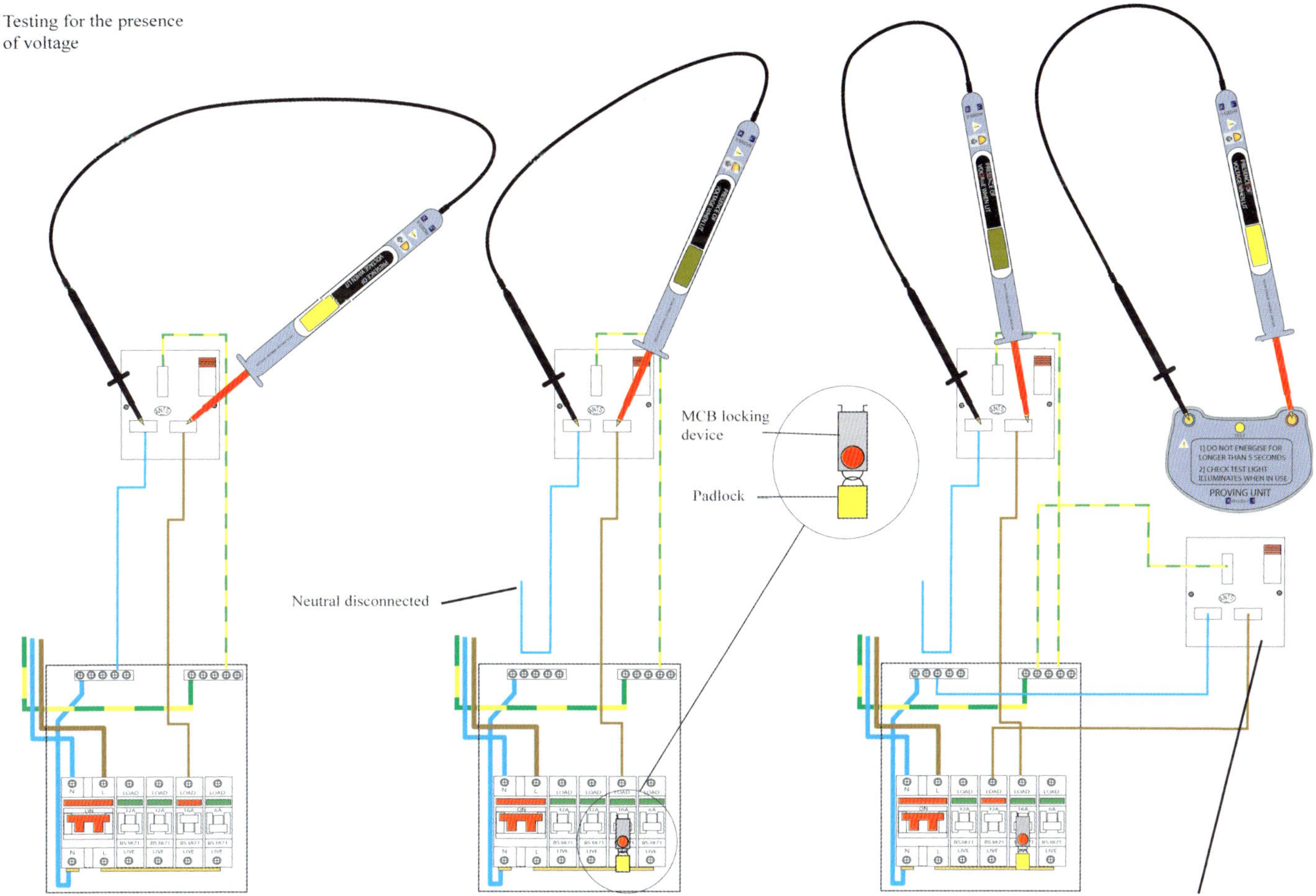

Fig: 1A

Fig: 1B

Fig: 1C

INITIAL VERIFICATION

The regulations state that before testing can commence, an initial verification inspection shall precede the metered testing and that the inspection shall be made to verify that the installed electrical equipment is:

- In compliance with British Standards and/or European Harmonisation Standards appropriate to the intended use of the installed equipment
- The installed equipment is correctly selected and erected in accordance with the regulations
- The installed equipment is not visibly damaged or defective so as to impair safety

and that the inspection shall include the checking of the following items where relevant to the installation and where necessary, during erection of the equipment. This means that some of the visual inspections can be carried out during erection of the equipment and therefore need not be re-tested.

- Connection of conductors
- Identification of conductors
- Routing of cables
- Selection of conductors
- Connection of single-pole devices
- Correct connection of equipment
- Presence of fire barriers
- Methods of protection against electric shock
- Prevention of detrimental influences
- Presence of appropriate devices for isolation and switching
- Presence of under-voltage protective devices
- Choice of setting of protective devices
- Labelling of protective devices, switches and terminals
- Selection of equipment appropriate to external influences
- Adequacy of access to switchgear and equipment
- Presence of danger notices and other warning signs
- Presence of diagrams, instructions and similar information
- Erection methods

Remember, if any of the initial verification checks require you to remove covers then you will need to carry out safe isolation, otherwise you will contravene the Electricity at Work Act 1989.

The key point with all electrical work is that you maintain your and everyone's safety when carrying out such work.

TESTING YOUR METER

It's no good just unpacking your meters and using them. It is very important that you regularly check your instruments to make sure they are in good and safe working order.

You must make sure that your instruments have a current calibration certificate, otherwise any tests made and recorded could be void.

Depicted in Fig: 2 page 9 you will see that there are a few tests that should be carried out to verify that the instrument is safe to use.

The sequence is as follows:

- Check meter for damage
- Check batteries are in good condition
- Check that the leads and probes are not damaged
- Check that the probes are of the appropriate type (GS38)
- Carry out open circuit test on all ranges to prove function of meter
- Carry out closed circuit test on all ranges
- Zero instrument on ohms scale

However, the sequence above only tests the scales of the instrument and test leads to their maximum and minimum values; it might be advisable to have various known value resisters to check the actual reading of the instrument on the Ωs and MΩ ranges, as seen in Fig: 2.

As you know, these instruments are very costly to replace so it is important that they are maintained in good working order and that they are looked after.

If a meter is dropped or inadvertently connected to a live supply it should be thoroughly checked and tested, if however you are in doubt about the accuracy of your instruments you can have them re-calibrated.

Note: before using any of your instruments make sure that all test probes and conductors to be tested are scrupulously clean to avoid incorrect test results.

Testing your meter

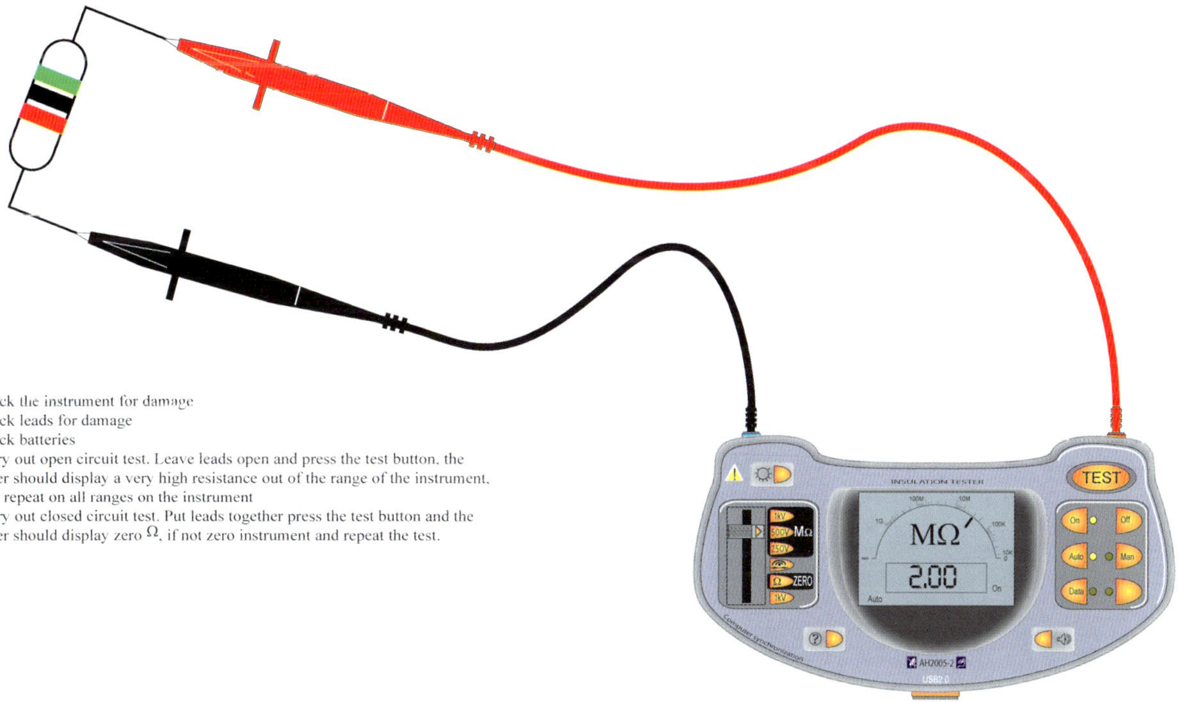

1] Check the instrument for damage
2] Check leads for damage
3] Check batteries
4] Carry out open circuit test. Leave leads open and press the test button, the meter should display a very high resistance out of the range of the instrument, now repeat on all ranges on the instrument
5] Carry out closed circuit test. Put leads together press the test button and the meter should display zero Ω, if not zero instrument and repeat the test.

Fig: 2

TEST 1
CONTINUITY OF PROTECTIVE CONDUCTORS INCLUDING MAIN AND SUPPLEMENTARY EQUIPOTENTIAL BONDING

CONTINUITY OF PROTECTIVE CONDUCTORS IN 3 PLATE 2 WAY LIGHTING CIRCUIT

The use of links when testing rather than long test leads will prevent hazards and save time with calculations.

The reason for checking the continuity is to make sure that the protective conductors are continuous throughout their length and establish a value for R2.

Depicted in Fig: 3 on page 11 is an example of measuring the continuity of the protective conductor in a 3 plate 2 way lighting circuit (conversion method).

The sequence is as follows:

- Remove all loads
- Ensure Isolation
- Remove distribution board cover
- Bridge phase to cpc (circuit protective conductor) at the distribution board
- Select an Insulation continuity tester and set to measure Ws, zero instrument
- Remove the last two way switch and place instrument as shown
- Take reading

Because you have linked the phase to the cpc in this test, you will measure the resistance of the phase and cpc. You can see in the drawing that the phase and cpc are of the same length, so in order to arrive at a resistance for the cpc only, you will have to divide the reading by 2. **Remember that this is only true if the cpc is the same csa (cross sectional area) as the phase conductor (see also note 2 page 12).**

Example

Assuming that you have established a reading of 0.55Ω, to establish the resistance of the cpc divide the reading 0.55Ω by 2 = 0.275Ω. This is the actual resistance of the cpc and can be entered on the test documentation (point Q on the record of test results page 64).

Remember to remove all links before reinstating the supply.

2 way lighting circuit
Test: Continuity of the CPC
Reason: To verify that the CPC is continuous throughout it's length

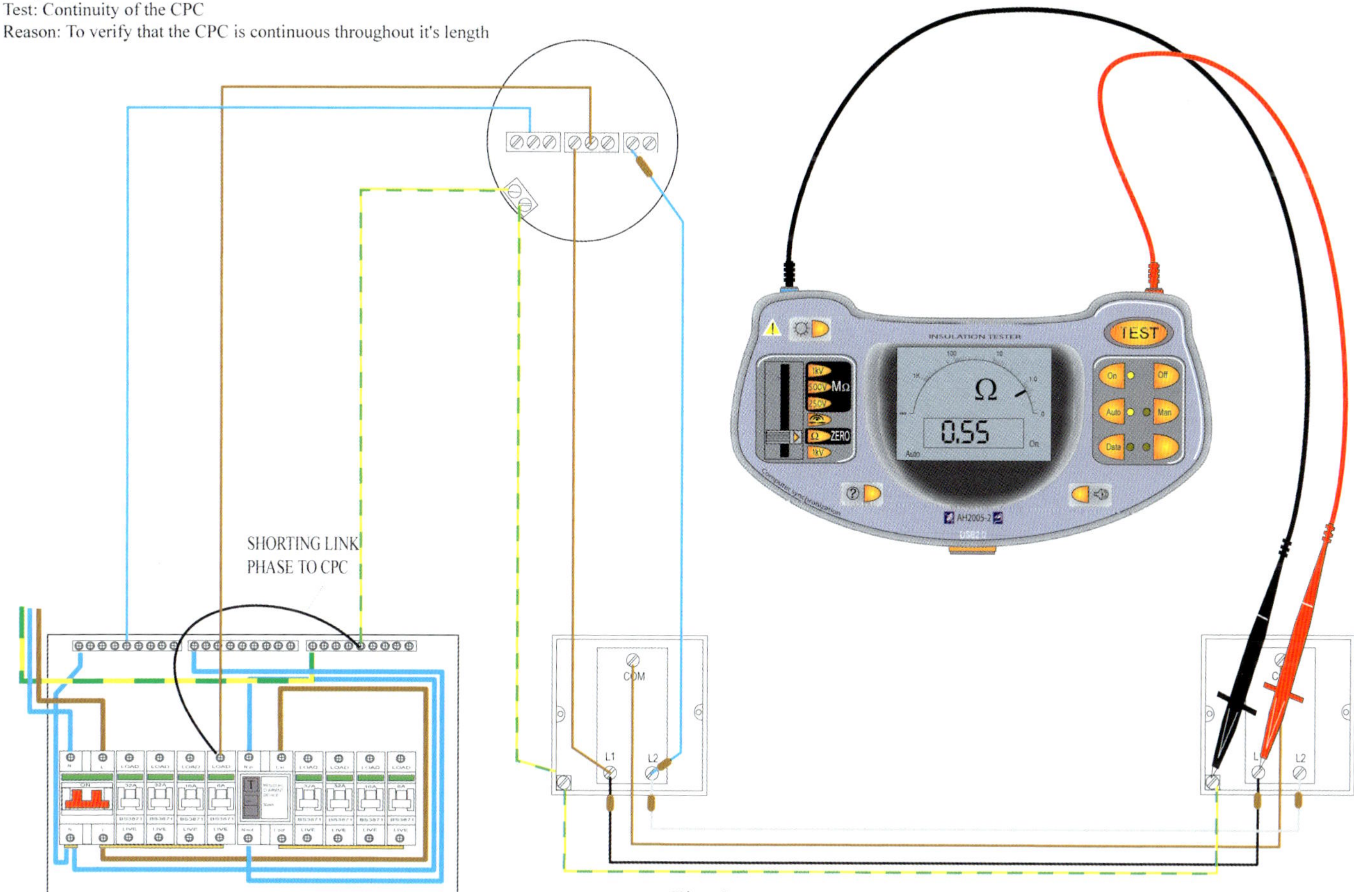

Customer control unit split board

Fig: 3

CONTINUITY OF PROTECTIVE CONDUCTORS IN 3 PLATE 1 WAY LIGHTING CIRCUIT

Depicted in Fig: 4 on page 13 is an example of measuring continuity of the protective conductor in a 3 plate 1 way lighting circuit.

The sequence is as follows:

- Remove all loads
- Ensure Isolation
- Remove distribution board cover
- Bridge phase to cpc (circuit protective conductor) at the distribution board
- Select an Insulation continuity tester and set to measure Ωs, zero instrument
- Remove light switch and place instrument as shown
- Take reading

Because you have linked the phase to the cpc in this test, you will measure the resistance of the phase and cpc. You can see in the drawing that the phase and cpc are of the same length, so in order to arrive at a resistance for the cpc only, you will have to divide the reading by 2. **Remember that this is only true if the cpc is the same csa (cross sectional area) as the phase conductor.**

<u>Example</u>
Assuming that you have established a reading of 0.4Ω, to establish the resistance of the cpc divide the reading 0.4Ω by 2 = 0.2Ω. This is the actual resistance of the cpc and can be entered on the test documentation (point Q on the record of test results page 64).

Note 2: These tests must be taken at the last point that the cpc terminates in the circuit, you only need to measure and record the maximum length of the cpc, but it will be necessary to verify that all the cpcs terminated in other switches before this point are continuous from the ceiling outlets to the switches.

Remember to remove all links before reinstating the supply.

1 way lighting circuit
Test: Continuity of the CPC
Reason: To verify that the CPC is continuous throughout it's length

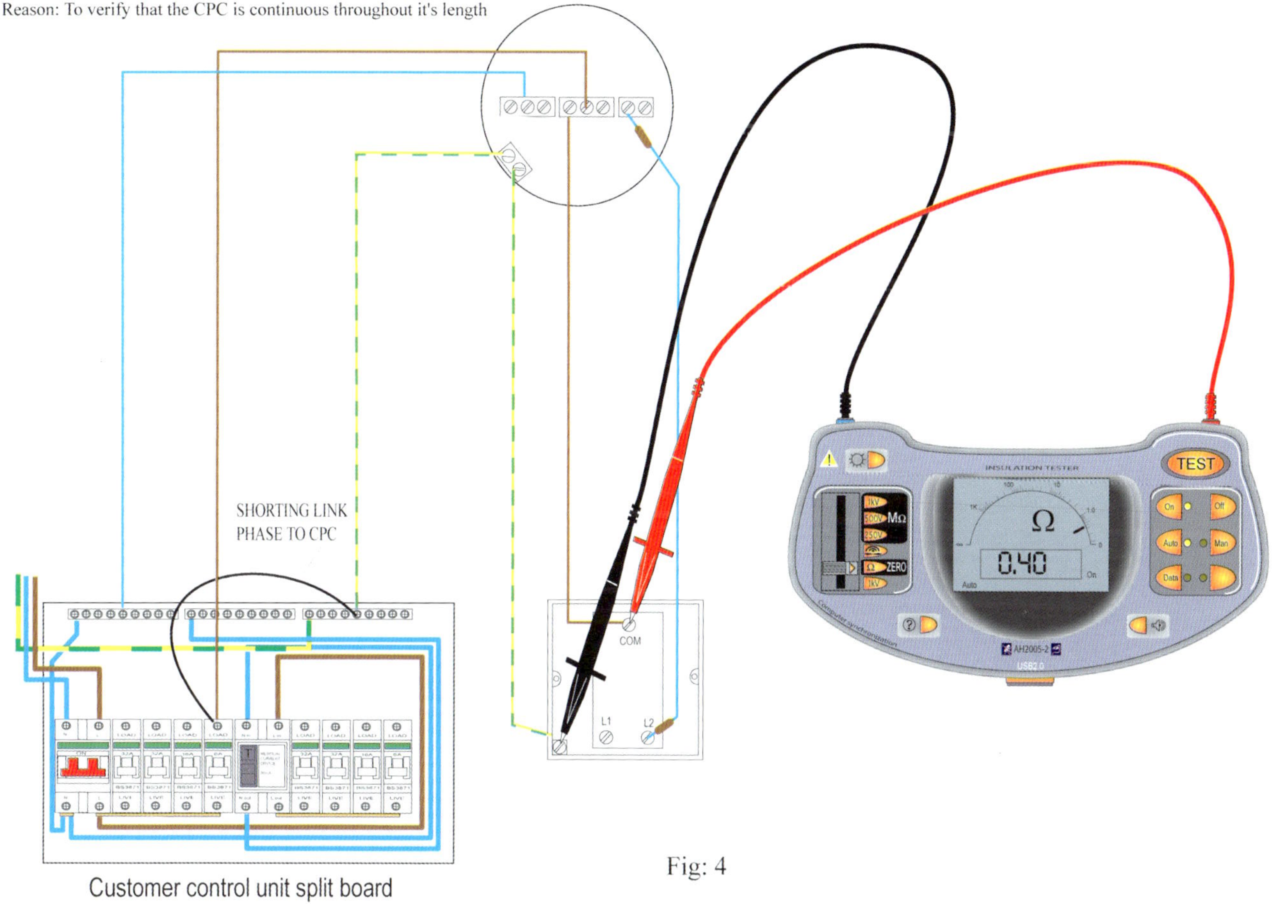

Fig: 4

CONTINUITY OF PROTECTIVE CONDUCTORS IN RING FINAL CIRCUIT

Depicted in Fig: 5 on page 15 is an example of measuring continuity of the protective conductor in a ring final circuit.

The sequence is as follows:

- Remove all loads
- Ensure Isolation
- Remove distribution board cover
- Select an Insulation continuity tester and set to measure Ωs, zero instrument
- Remove conductors for circuit to be tested
- Place instrument on the circuit as shown
- Take reading and record at point Q on record of test results page 64

You have established the actual resistance of the cpc. In this instance there will be no need to adjust the reading as we did for the lighting circuits.

During earth faults in ring final circuits, the ring is in effect placed in parallel during the fault and therefore the resistance will be a reciprocal of the phase and cpc resistance. Do not worry too much about this now, I will explain in detail later when confirming continuity of ring final circuit conductors.

EXAMPLE

Assume that you have established a reading of 0.8Ω, this is the actual resistance of the cpc in the ring final circuit. You can see already one of the benefits of ring circuits, they tend to make testing a lot easier, by this I mean you do not have to adjust the figures to arrive at the resistance for the cpc.

You could also multiply the reading obtained by 0.25 to give you the R2 value, i.e 0.8Ω x 0.25 = 0.2Ωs this is the R2 value. The same can also be carried out for the phase conductor which in turn will give you the R1 value.

Ring final circuits

Test: Continuity of the CPC

Reason: To verify that the CPC is continuous throughout it's length

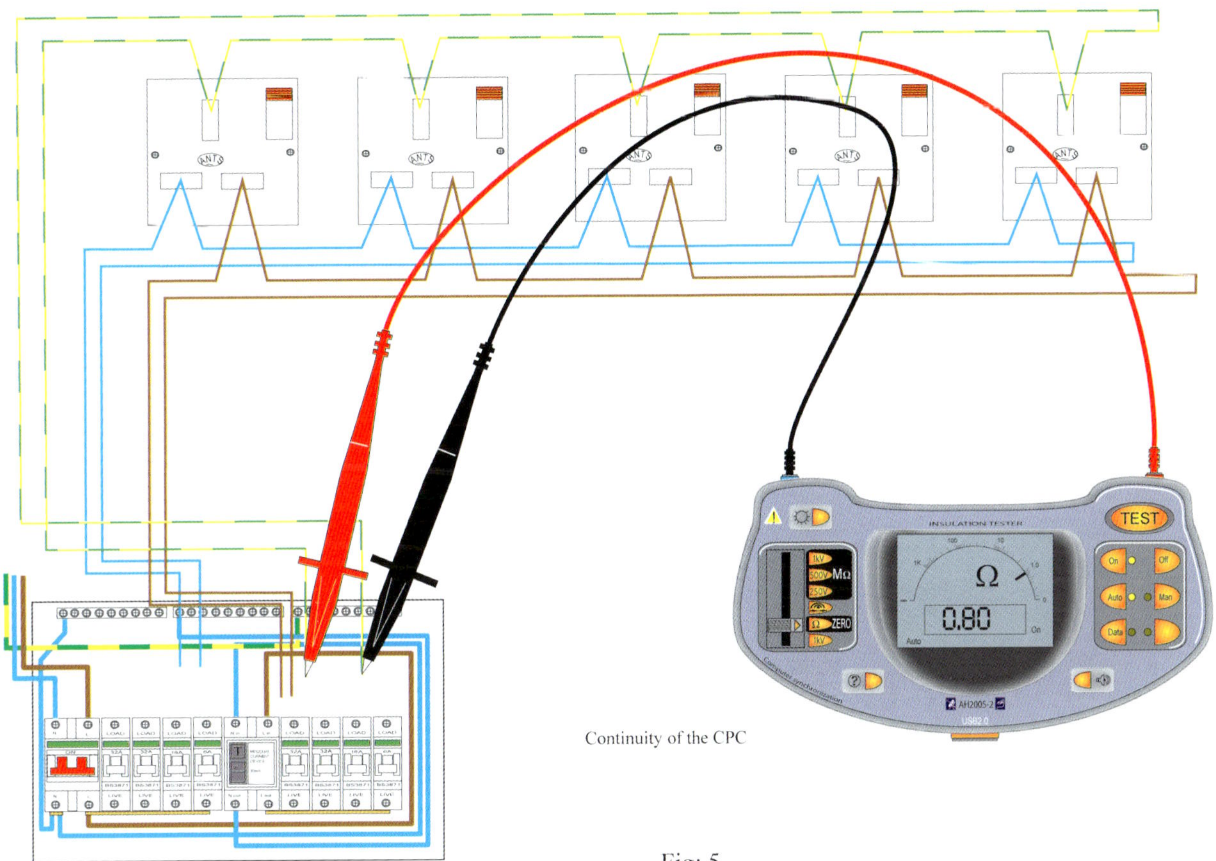

Customer control unit split board

Continuity of the CPC

Fig: 5

CONTINUITY OF PROTECTIVE CONDUCTORS IN 3 PHASE MOTOR CIRCUITS WIRED IN STEEL TRUNKING AND CONDUIT

Depicted in Fig: 6 on page 17 is an example of measuring continuity of the protective conductor in a 3 phase motor circuit.

The sequence is as follows:

- Remove all loads
- Ensure Isolation
- Remove distribution board cover or isolator for the circuit
- Select an Insulation continuity tester and set to measure Ωs, zero instrument
- Place links at the main isolator, local isolator and in the contactor as shown
- Place instrument on the circuit as shown
- Take reading and record

You have established the resistance of the cpc and the phase conductors. In this case the cpc is the steel conduit and trunking. You will need to establish what the resistance of the phase conductor is to be able to find the actual resistance of the cpc. Unfortunately it is not as simple as dividing the answer by 2 because the conduit and trunking will generally have a larger csa than the phase conductor.

EXAMPLE

Assume that you have established a reading of 0.3Ω, this is the resistance of the conduit trunking and phase conductor. Now carry out the test depicted in Fig: 7 page 19 referring to the text on page 18.

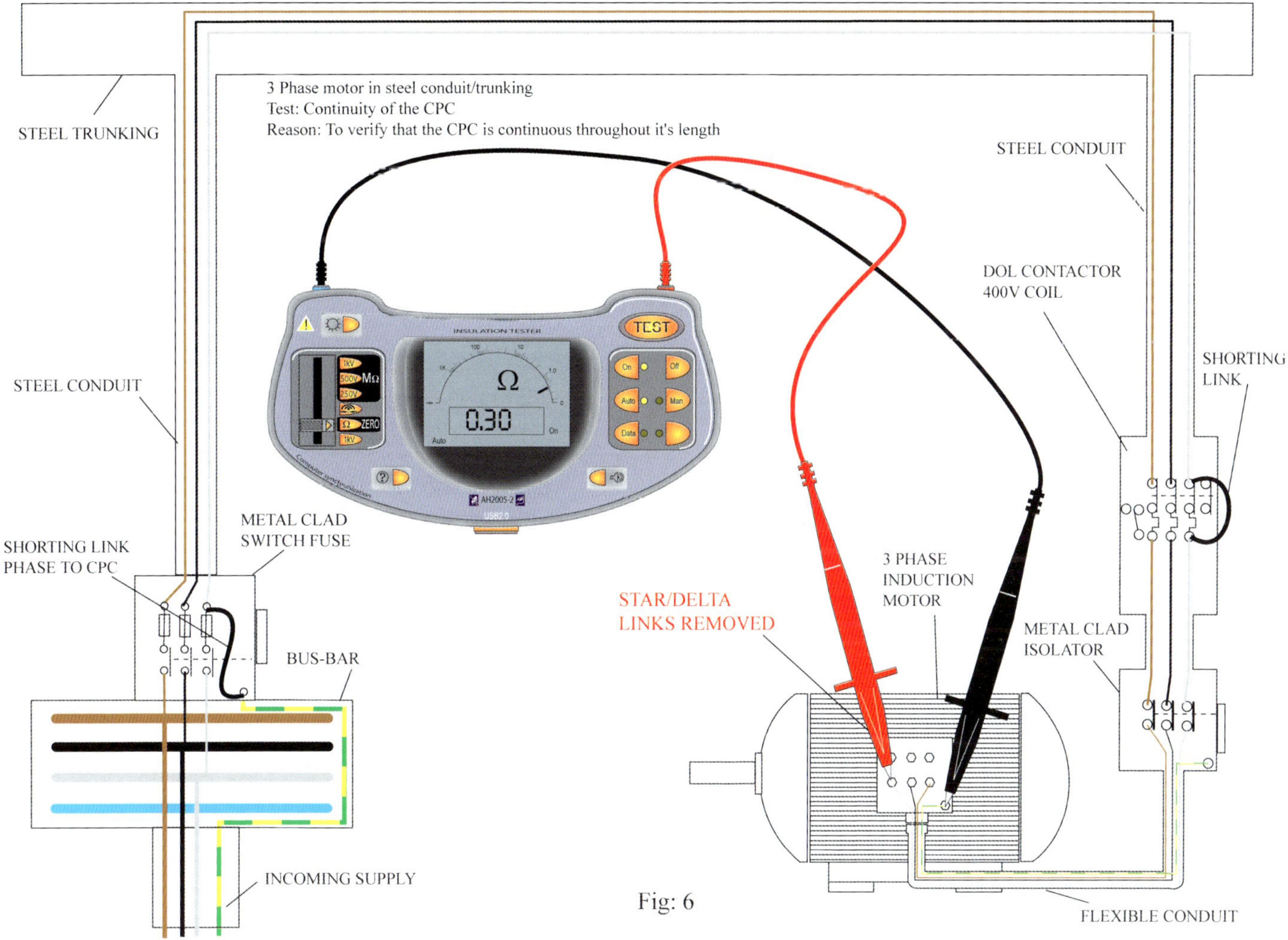
3 Phase motor in steel conduit/trunking
Test: Continuity of the CPC
Reason: To verify that the CPC is continuous throughout it's length
STEEL TRUNKING
STEEL CONDUIT
STEEL CONDUIT
DOL CONTACTOR
400V COIL
SHORTING
LINK
METAL CLAD
SWITCH FUSE
SHORTING LINK
PHASE TO CPC
BUS-BAR
INCOMING SUPPLY
STAR/DELTA
LINKS REMOVED
3 PHASE
INDUCTION
MOTOR
METAL CLAD
ISOLATOR
FLEXIBLE CONDUIT
TEST
0.30
INSULATION TESTER
AH2005-2

Fig: 6

Continuing from page 16

Whilst referring to Fig: 7 you will notice that the **star links** have been removed on the motor. The reason for this is that if these links are not removed you will measure the resistance of the motor windings.

In the first test you established a reading of 0.3Ω for the phase and cpc. You now need to carry out the test in Fig: 7 page 19.

Assume that you have established a reading of 0.4Ω, this is the resistance of 2 phase conductors either red and yellow, red and blue or blue and yellow.

If you measure the resistance of all the pairs of conductors, e.g. you have three sets of values, you can compare them to each other and they should be substantially the same reading.

As the conductors are the same length and csa we can divide the figure by 2, thus 0.4Ω divided by 2 gives us a reading of 0.2Ω, which is the resistance of one of the phase conductors.

Now subtract this from the original reading of 0.3Ω which will give a calculated value of 0.3Ω - 0.2Ω = 0.1Ω which is the actual resistance of the conduit and trunking acting as the cpc (record at point Q on record of test results page 64).

You can see from the resistance values that the trunking and conduit are of a greater csa than the conductors, due to the fact that the resistance is lower for the trunking and conduit than that of the phase conductor.

It is important to establish the resistance of the trunking and conduit, because the trunking and conduit will carry any earth fault currents, and if the resistance is not low enough then disconnection of the circuit may not be achieved within the time allowed. However, sometimes the resistance might appear to be too high due to badly erected systems or loose couplers and adapters, and these would need to be rectified and re-tested before commencing with further tests.

Remember to remove all links before reinstating the supply.

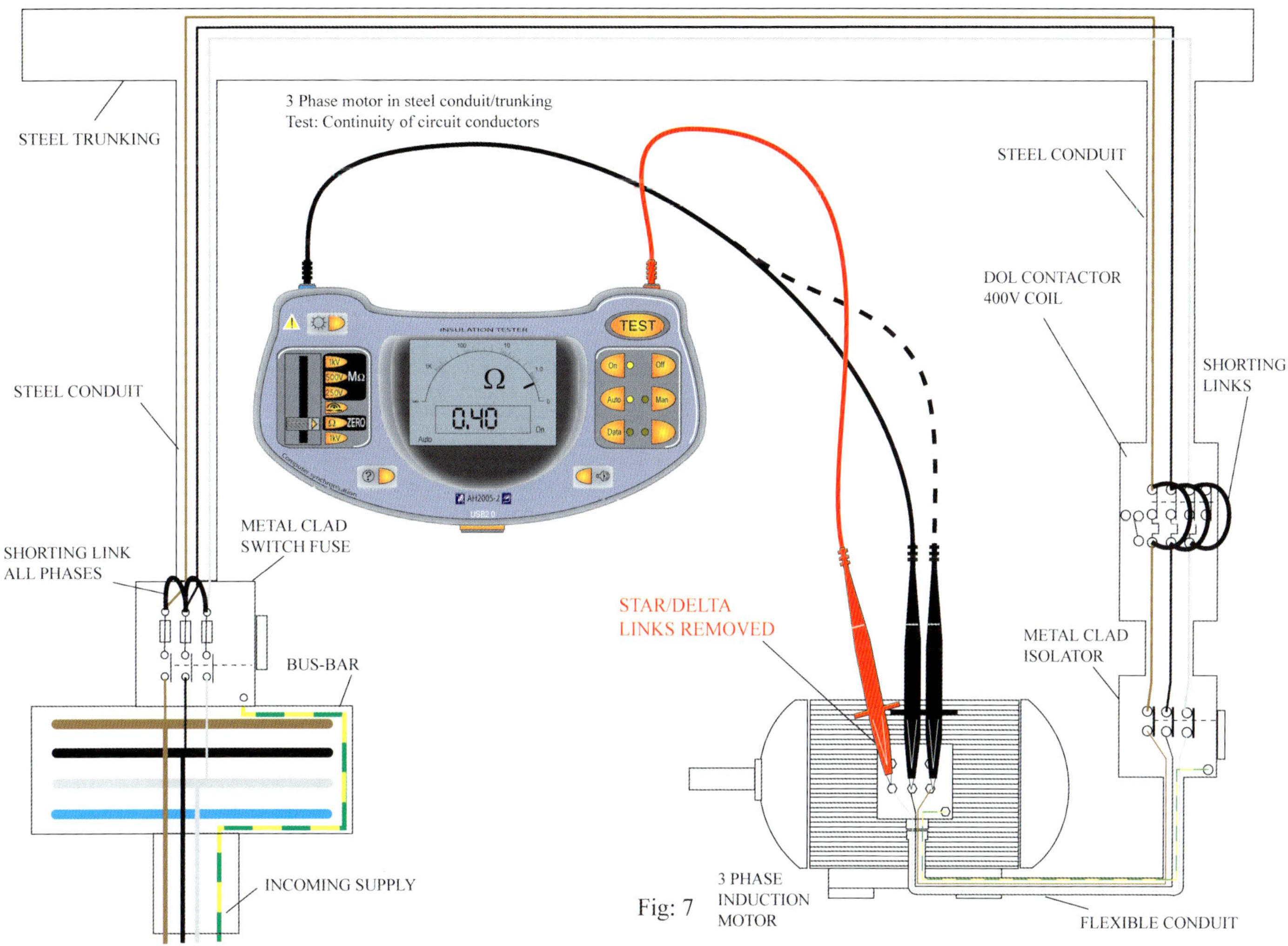
3 Phase motor in steel conduit/trunking
Test: Continuity of circuit conductors
STEEL TRUNKING
STEEL CONDUIT
SHORTING LINK
ALL PHASES
METAL CLAD
SWITCH FUSE
BUS-BAR
INCOMING SUPPLY
TEST
0.40
STEEL CONDUIT
DOL CONTACTOR
400V COIL
SHORTING
LINKS
STAR/DELTA
LINKS REMOVED
METAL CLAD
ISOLATOR
3 PHASE
INDUCTION
MOTOR
FLEXIBLE CONDUIT

Fig: 7

CONTINUITY OF MAIN EQUIPOTENTIAL BONDING

Depicted in Fig: 8 on page 21 is an example of measuring continuity of the main equipotential bonding conductors.

This is one of the occasions that I can condone the use of long test leads; generally this is the easiest way of carrying out the test, although you will have to bear in mind the dangers of someone tripping over the long test leads.

The sequence is as follows:

- Remove all loads
- Ensure Isolation
- Select an Insulation continuity tester and set to measure Ωs, zero instrument
- Place instrument on the circuit as shown
- Take reading and record at point P on record of test results page 64

Make sure that each main equipotential bonding conductor under test is disconnected from the main earth terminal, or at the bonding clamp end, to avoid parallel paths through other bonding conductors.

Isolation is a precaution that needs to be carried out because if an earth fault were to occur whilst carrying out the test then a potential (voltage) could be introduced on to the main bonding and give rise to the risk of shock. Bonding conductors can be terminated at the earth bar in the distribution board.

Note: If isolation cannot be achieved then you should **not** disconnect any bonding conductors.

Also remember that if you are using long test leads, make sure that you zero out the resistance of the leads or at least deduct the resistance of the leads from your measured result.

There is no absolute reading or table that you can reference to see if the resistance of the bonding is correct. All you should expect is that the resistance is negligible or less than 0.05Ωs.

Main Equipotential Bonding
Test: Continuity of the main equipotential bonding conductors

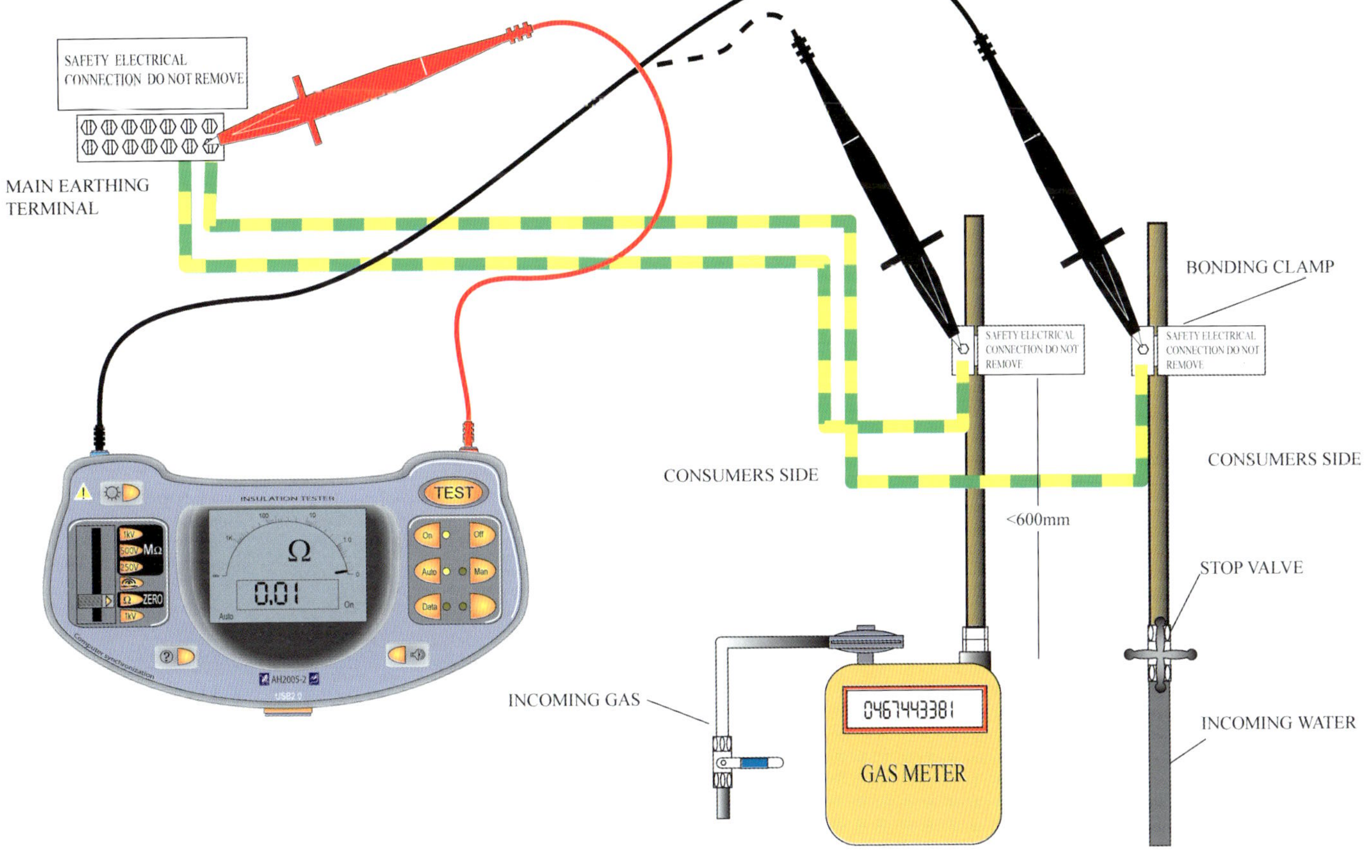

Fig: 8

CONTINUITY OF SUPPLEMENTARY BONDING CONDUCTORS

Depicted in Fig: 9 on page 23 is an example of measuring continuity of the supplementary bonding conductors.

This is another occasion that I can condone the use of long test leads; generally this is the easiest way of carrying out the test, although you will have to bear in mind the dangers of someone tripping over the long test leads.

The sequence is as follows:

- Remove all loads
- Ensure Isolation
- Select an Insulation continuity tester and set to measure Ωs, zero instrument
- Place instrument on the circuit as shown
- Take reading and record (position P on test sheet, page 64)

Note: Isolation is only a precaution, because if an earth fault were to occur whilst carrying out the test, a potential (voltage) could be introduced onto the main bonding and give rise to the risk of shock. Bonding conductors can be terminated at the earthing bar in the distribution board.

Also remember that if you are using long test leads make sure that you zero out the resistance of the leads or at least deduct the resistance of the leads from your measured result.

There is no absolute reading or table that you can reference to see if the resistance of the bonding is correct. All you should expect is that the resistance is negligible or extremely small.

Make sure that each supplementary bonding conductor under test is disconnected from the bonding clamp, to avoid parallel paths through other bonding conductors.

Supplementary bonding

Test: Continuity of the supplementary bonding conductors

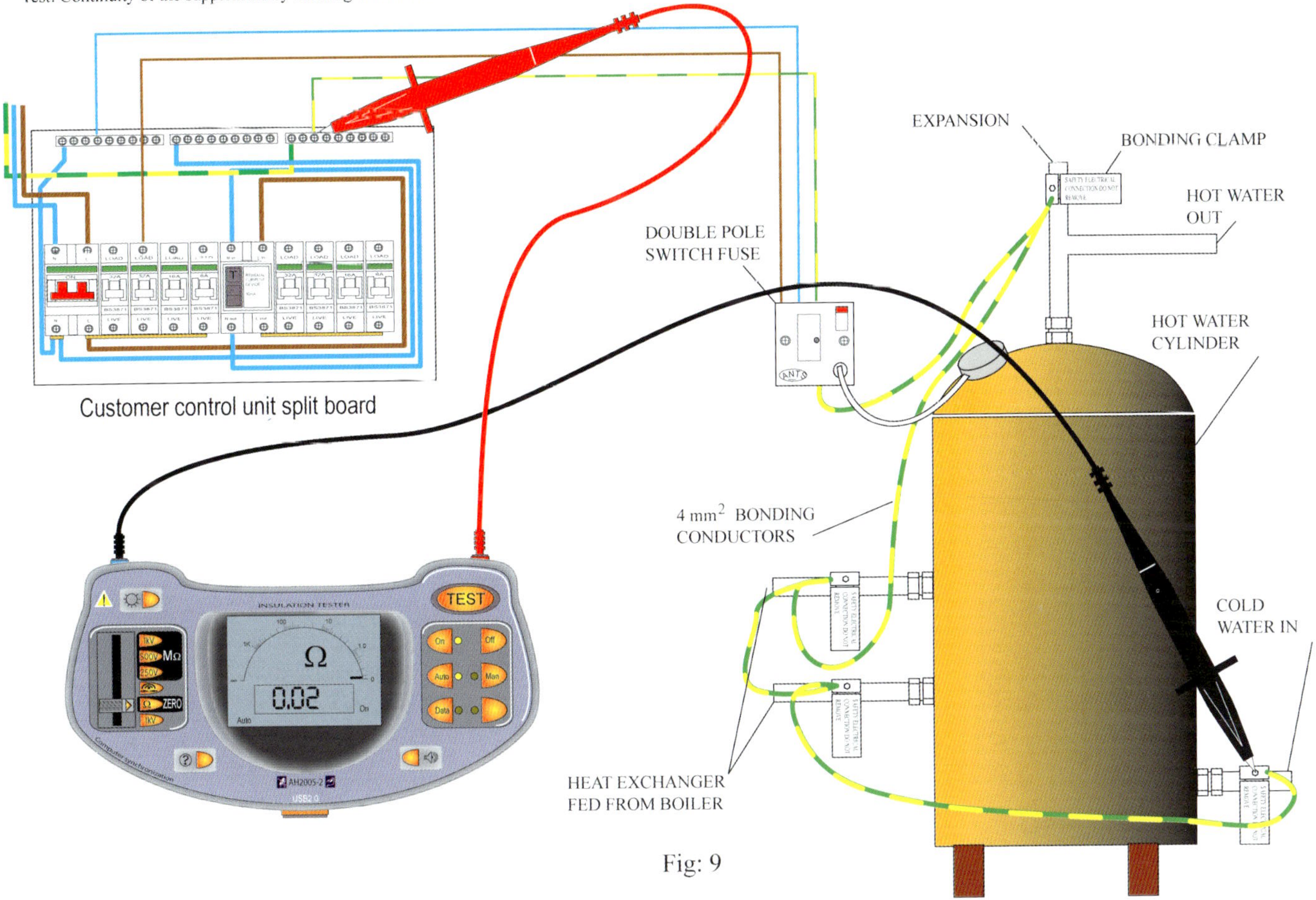

Fig: 9

TEST 2
CONTINUITY OF RING FINAL CIRCUIT CONDUCTORS

This is one of the most complicated tests to understand because not only are we concerned with measuring the continuity but also proving that ring final circuits are connected as a ring and that no interconnections exist.

Depicted in Fig: 10 on page 25 is an example of measuring continuity of the phase and neutral conductors in ring final circuits.

The sequence is as follows:

- Remove all loads
- Ensure Isolation
- Disconnect circuit conductors
- Select an Insulation continuity tester and set to measure Ωs, zero instrument
- Place instrument on the circuit as shown in Fig: 10 on page 25
- Take readings and record

This test will establish that the resistance of the phase and neutral conductors are substantially the same.

Example

Assume that you have measured the resistance of the phase conductor and found the resistance to be 0.4Ω. When you measure the resistance of the neutral conductor the reading should be substantially the same as that of the phase conductor e.g. 0.4Ω.

If these readings are not substantially the same then there may be a problem with the circuit, such as loose terminals and or conductors terminated incorrectly.

You could also multiply the reading obtained by 0.25 to give you the R1 value,
i.e 0.4Ω x 0.25 = 0.1Ωs this is the R1 value.

Remember that you have already carried out continuity of cpc in the ring final circuit so we will be needing to add this to calculations later on.

There are three further tests to complete the sequence refer to page 26

Ring final circuits
Test: Continuity of phase and neutral conductors
Reason: To verify that the phase and neutral conductors are continuous throughout their length

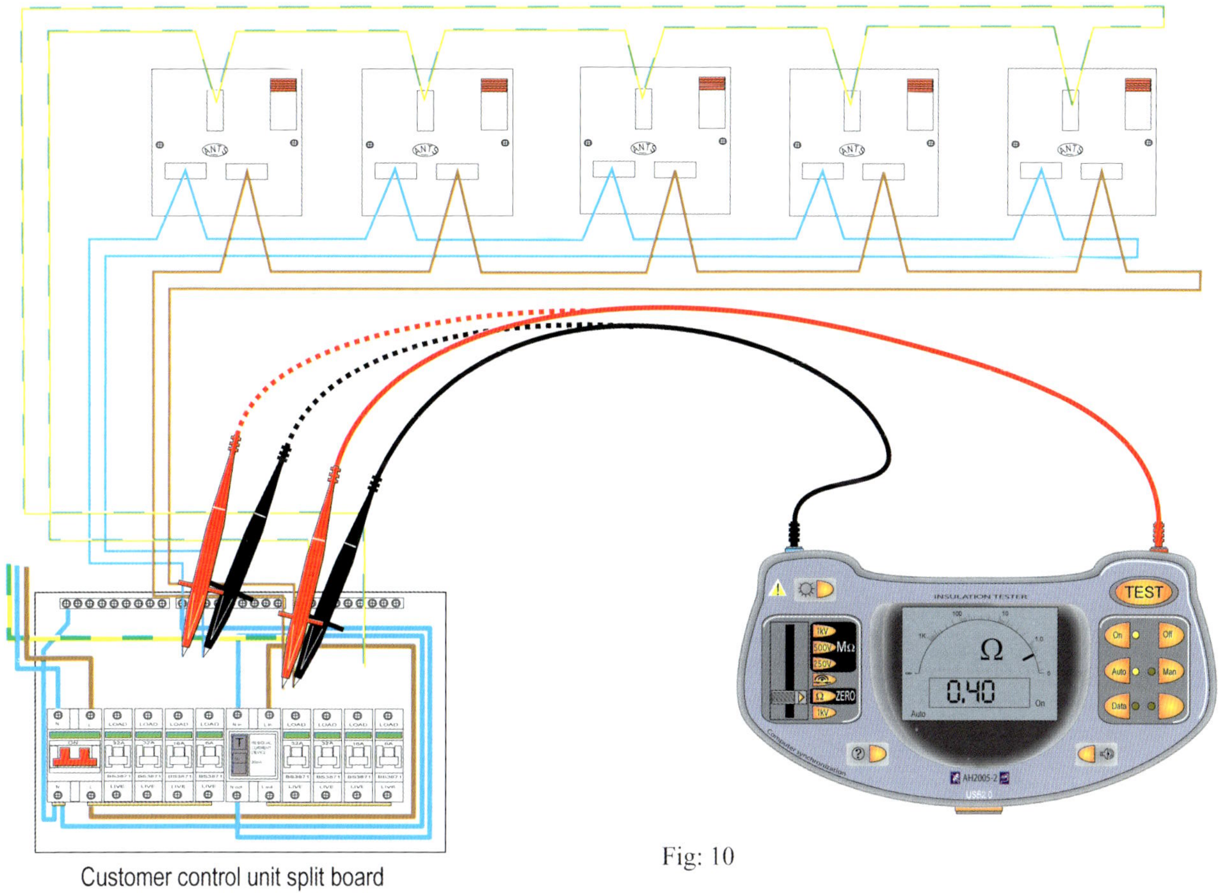

Fig: 10

After establishing the resistance of the phase and neutral conductors we can carry on with the further tests to confirm that the ring is indeed wired as a ring.

Now refer to Fig: 11A on page 27. You can see that a link has been inserted connecting the outgoing phase to the incoming neutral conductor. If you established a resistance of 0.4Ω for each of the phase and neutral conductors previously, then when connected in this fashion we should have twice the resistance e.g. phase resistance + neutral resistance = 0.4Ω + 0.4Ω = 0.8Ω.

Now refer to Fig: 11B on page 27. You can see that a further link has been inserted connecting the incoming phase to the outgoing neutral conductors; all four ends of the ring are now bridged. Having established a resistance of 0.8Ω for the previous test, when connected in this fashion we should have a quarter of the previous resistance, i.e. 0.25 X (phase resistance + neutral resistance) = 0.25 X (0.4Ω + 0.4Ω) = 0.2Ω.

With this test result of 0.2Ω we can test at each socket on the ring circuit see Fig: 11C. All of the sockets on the ring should have substantially the same reading of 0.2Ω. If this reading is not the same, then there may be interconnections in the ring which need to be rectified or you will need to reconfirm that the outgoing and incoming conductors have been clearly identified.

You can now repeat the tests substituting the cpc for the neutral conductor. This will give you the (R1 + R2) value for the ring final circuit (record at point U on record of test results page 64).

If you know or have established the external impedance Ze then adding the two together will give you the value for earth loop impedance.

As this is a fairly difficult test to get to grips with I suggest that you practice the tests on a ring circuit at length.

Remember to remove all links before reinstating the supply.

Ring final circuits
Test: Confirmation of the ring final circuit
Reason: To verify that the ring is wired in a ring and that no interconnections exist in the ring

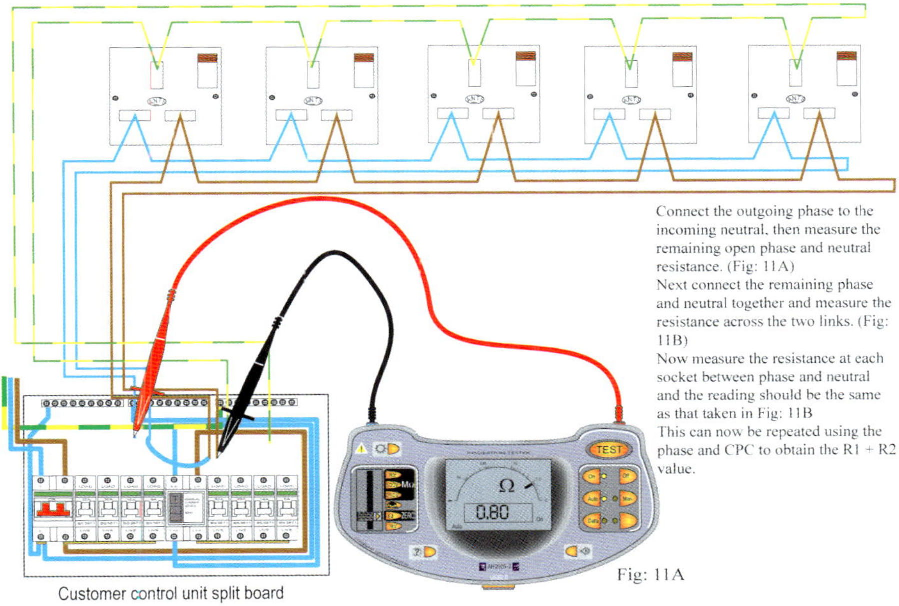

Connect the outgoing phase to the incoming neutral, then measure the remaining open phase and neutral resistance. (Fig: 11A)
Next connect the remaining phase and neutral together and measure the resistance across the two links. (Fig: 11B)
Now measure the resistance at each socket between phase and neutral and the reading should be the same as that taken in Fig: 11B
This can now be repeated using the phase and CPC to obtain the R1 + R2 value.

Fig: 11A

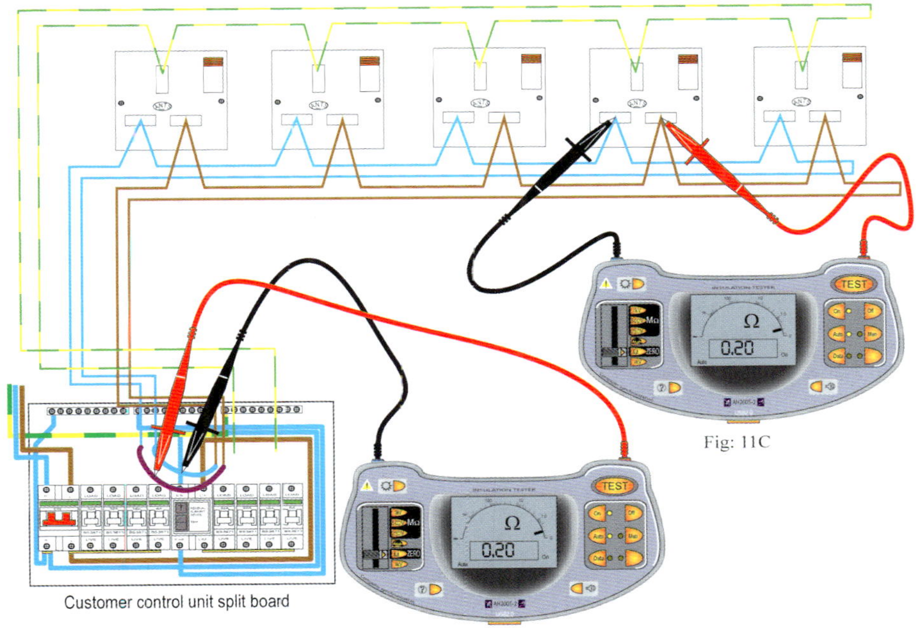

Fig: 11C

Fig: 11B

CONTINUITY OF CIRCUIT CONDUCTORS IN LIGHTING CIRCUITS

Even though this is not a laid down test in the regulations you will need to carry out the test to establish that the circuit functions correctly. These tests will save time during the commissioning phase of a new installation. The time savings will be made by highlighting switches installed upside down, 2 way lighting circuits incorrectly wired and also helps towards achieving polarity tests.

Depicted in Fig: 12 on page 29 is an example of checking continuity of the circuit conductors in a 3 plate 1 way lighting circuit.

The sequence is as follows:

- Remove all loads
- Ensure Isolation
- Place links on the circuit as indicated in Fig: 12
- Select an Insulation continuity tester and set to measure Ωs, zero instrument
- Place instrument on the circuit as shown
- Take reading and record

Note: If the switch is turned on and off with the meter connected as shown the meter will signify the turning on and off of the light.

Remember to remove all links before reinstating the supply.

Lighting circuit

Test: Continuity of circuit conductors

Reason: To verify that the circuit functions correctly and once a reading is obtained it can be used to give an approximate resistance of the phase conductor R1.

Note: The reading is approximate due to the added resistance of the switch line.

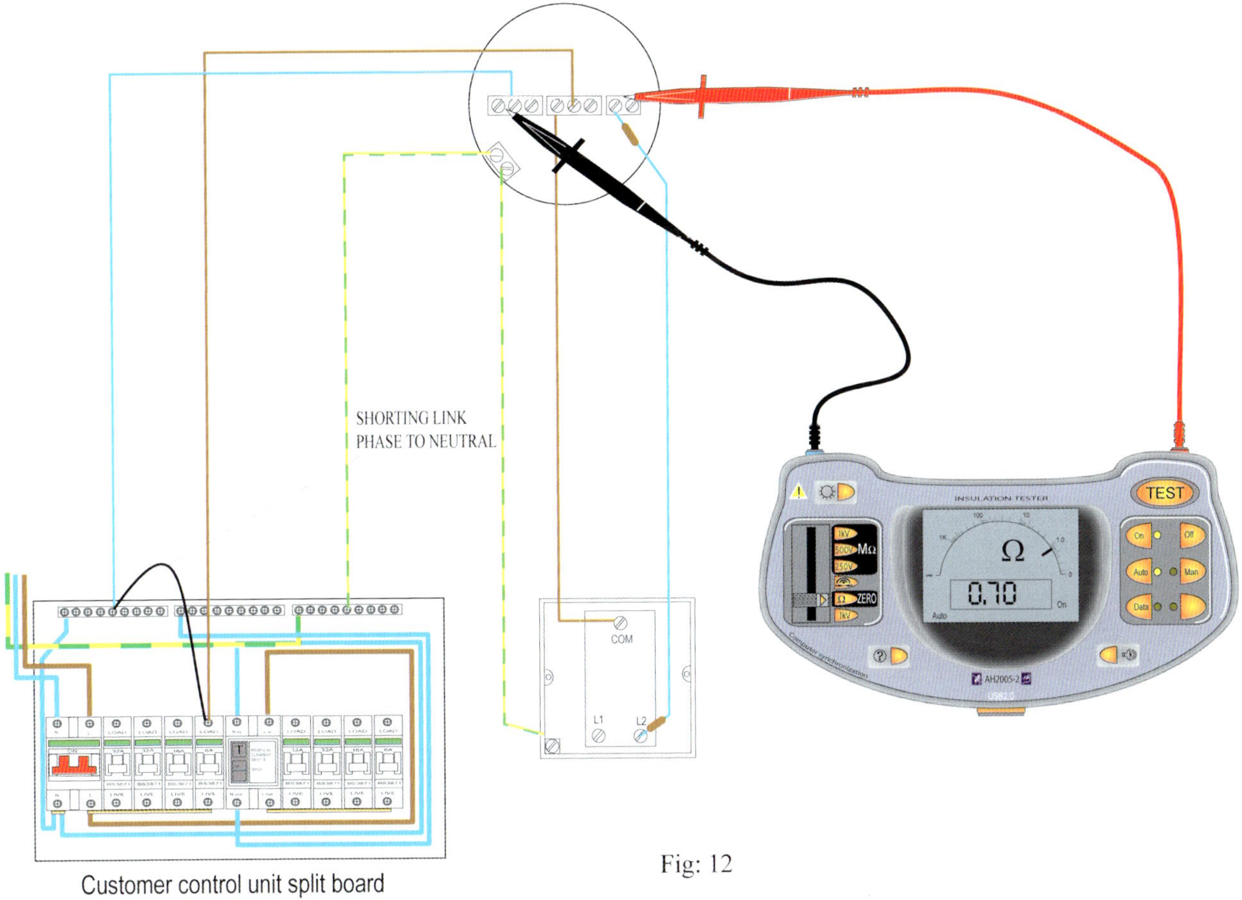

Fig: 12

TEST 3
INSULATION RESISTANCE IN LIGHTING CIRCUITS

Depicted in Fig: 13 on page 31 is an example of checking Insulation resistance in a 3 plate 1 way lighting circuit.

All loads must be disconnected and switches turned on. Remember to remove any voltage sensitive electronic devices in the circuit such as dimmer switches, timers, controllers, electronic starters, and controlgear for discharge lamps to prevent damage and misleading results.

The sequence is as follows:

- Remove all loads
- Ensure Isolation
- Select an Insulation continuity tester and set to 500V scale
- Place instrument on the circuit as shown
- Take reading and record at points W, X, Y on record of test results page 65

The reason for carrying out the insulation resistance test, is to check that the insulation of the circuit conductors has not broken down. With basic theory if you apply a potential to a conductor electrons will flow in the circuit, that is exactly what you are trying to do. If you apply a voltage with the test instrument, you are trying to encourage electron flow across the insulation of the conductors. If electrons flow then there is a breakdown in the insulation and the resistance can be measured.

This table gives the minimum values allowed

CIRCUIT NOMINAL VOLTAGE (V)	TEST VOLTAGE d.c. (V)	MINIMUM INSULATION RESISTANCE (MΩ)
SELV AND PELV *	250	0.25
UP TO AND INCLUDING 500V WITH THE EXCEPTION OF THE ABOVE SUPPLIES	500	0.5
ABOVE 500V	1000	1.0

*SELV = Separated Extra-low Voltage PELV = Protected Extra-low Voltage

The IEE Guidance notes on testing state that if a reading of between that stated in the table and 2MΩ is obtained then further investigation is required.

Lighting circuit
Test: Insulation Resistance
Reason: Check for damage to cables and breakdown in insulation.
Test to be carried out between
- Phase & Neutral
- Phase & CPC
- Neutral & CPC

Instrument set to 500V

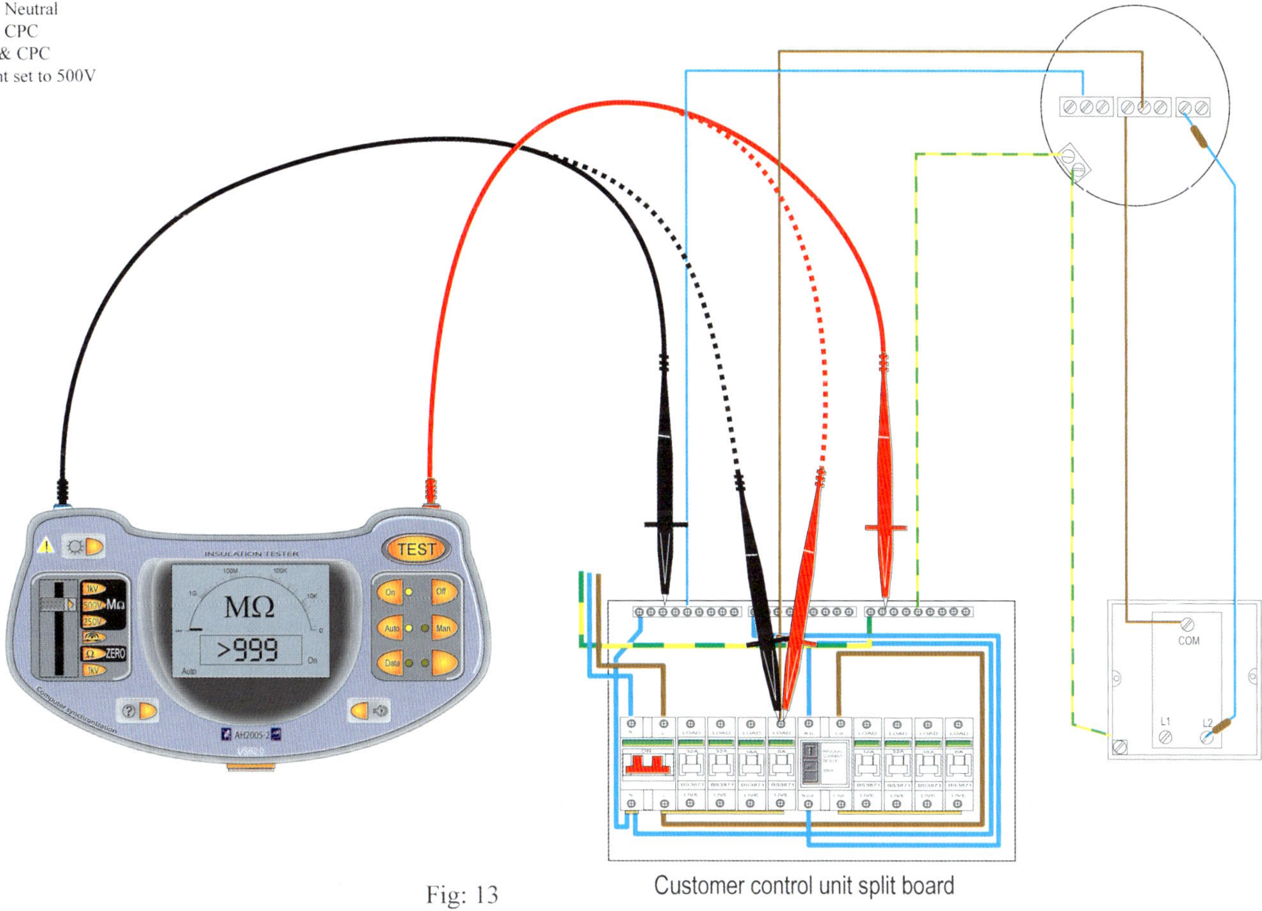

Fig: 13

INSULATION RESISTANCE IN RING FINAL CIRCUITS

Depicted in Fig: 14 on page 33 is an example of checking Insulation resistance in a ring final circuit.

All loads must be disconnected and switches turned on. Remember to remove any voltage sensitive electronic devices in the circuit such as dimmer switches, timers, controllers, electronic starters and controlgear for discharge lamps to prevent damage and misleading test results.

The sequence is as follows:

- Remove all loads
- Ensure Isolation
- Select an Insulation continuity tester and set to 500V scale
- Place instrument on the circuit as shown
- Take reading and record at points W, X, Y on record of test results page 65

The reason for carrying out insulation resistance test is to check that the insulation of the circuit conductors has not broken down. With basic theory if you apply a potential to a conductor electrons will flow in the circuit, that is exactly what you are trying to do. If you apply a voltage with the test instrument you are trying to encourage electron flow across the insulation of the conductors. If electrons flow then there is a breakdown in the insulation and the resistance can be measured.

This table gives the minimum values allowed

CIRCUIT NOMINAL VOLTAGE (V)	TEST VOLTAGE d.c. (V)	MINIMUM INSULATION RESISTANCE (MΩ)
SELV AND PELV *	250	0.25
UP TO AND INCLUDING 500V WITH THE EXCEPTION OF THE ABOVE SUPPLIES	500	0.5
ABOVE 500V	1000	1.0

*SELV = Separated Extra-low Voltage PELV = Protected Extra-low Voltage

The IEE Guidance notes on testing state that if a reading of between that stated in the table and 2MΩ is obtained then further investigation is required.

Ring final circuits
Test: Insulation resistance
Reason: To check for damage to cables and a breakdown in insulation.

Test to be carried out between
Phase & Neutral
Phase & CPC
Neutral & CPC
Instrument set to 500V

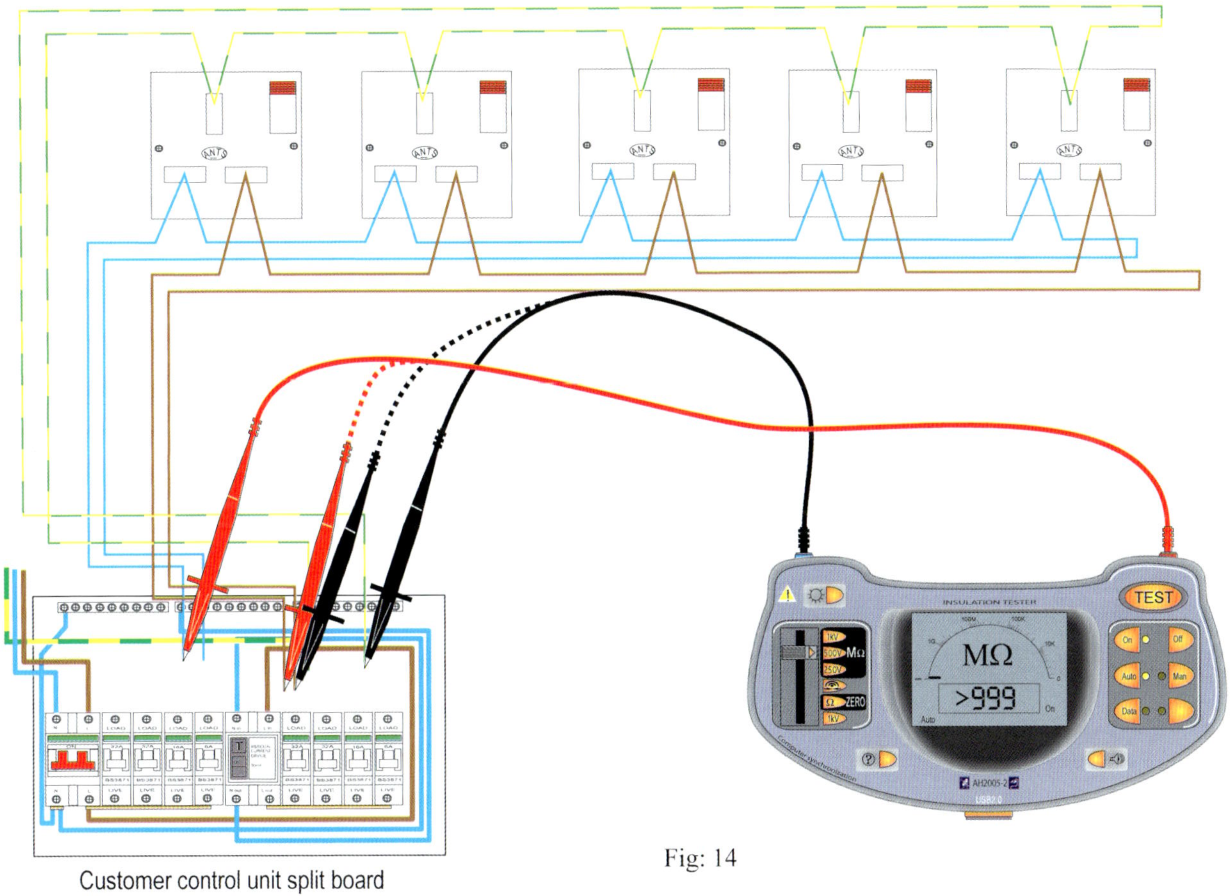

Fig: 14

INSULATION RESISTANCE IN 3 PHASE MOTOR CIRCUITS

Depicted in Fig: 15 on page 35 is an example of checking Insulation resistance in a 3 phase motor circuit.

All loads must be disconnected and switches turned on. Remember to remove any voltage sensitive electronic devices in the circuit such as dimmer switches, timers, controllers, electronic starters and controlgear for discharge lamps to prevent damage and misleading results. You will also have to disconnect the coil in the contactor and the star/delta links in the terminal box on the motor.

The sequence is as follows:

- Remove all loads
- Ensure Isolation
- Place links in circuit as indicated in Fig: 15
- Select an Insulation continuity tester and set to 500V scale
- Place instrument on the circuit as shown
- Test Br-Bl, Br-G, Bl-G, Br-E, Bl-E, G-E
- Take reading and record at points Z, AA, BB record of test results page 65

The reason for carrying out insulation resistance test is to check that the insulation of the circuit conductors has not broken down. With basic theory if you apply a potential to a conductor electrons will flow in the circuit, that is exactly what you are trying to do. If you apply a voltage with the test instrument you are trying to encourage electron flow across the insulation of the conductors. If electrons flow then there is a breakdown in the insulation and the resistance can be measured. **Remember to remove all links before reinstating the supply.**

This table gives the minimum values allowed

CIRCUIT NOMINAL VOLTAGE (V)	TEST VOLTAGE d.c. (V)	MINIMUM INSULATION RESISTANCE (MΩ)
SELV AND PELV *	250	0.25
UP TO AND INCLUDING 500V WITH THE EXCEPTION OF THE ABOVE SUPPLIES	500	0.5
ABOVE 500V	1000	1.0

*SELV = Separated Extra-low Voltage PELV = Protected Extra-low Voltage

The IEE Guidance notes on testing state that if a reading of between that stated in the table and 2MΩ is obtained then further investigation is required.

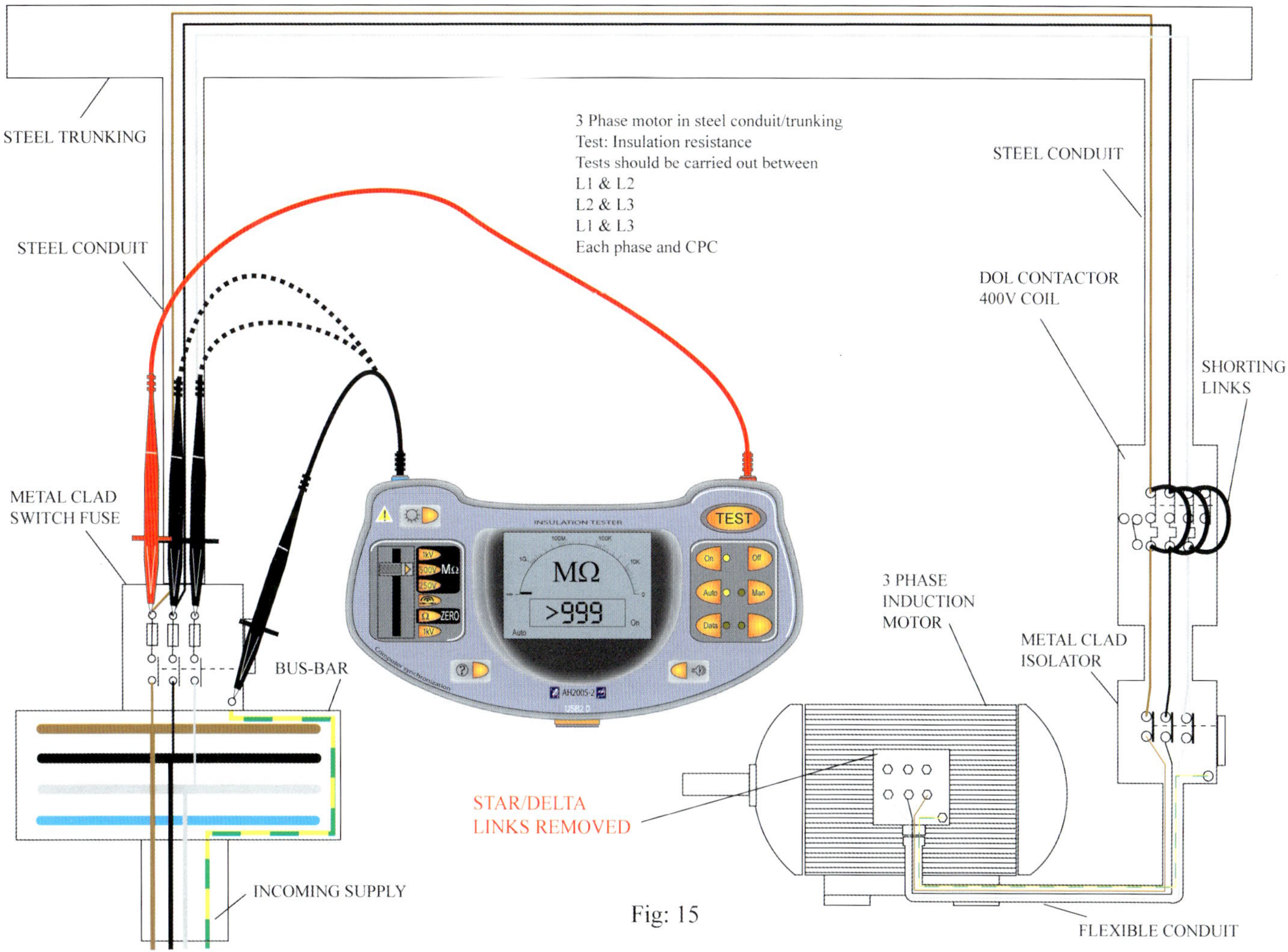
3 Phase motor in steel conduit/trunking
Test: Insulation resistance
Tests should be carried out between
L1 & L2
L2 & L3
L1 & L3
Each phase and CPC
STEEL TRUNKING
STEEL CONDUIT
METAL CLAD
SWITCH FUSE
BUS-BAR
INCOMING SUPPLY
STEEL CONDUIT
DOL CONTACTOR
400V COIL
SHORTING
LINKS
3 PHASE
INDUCTION
MOTOR
METAL CLAD
ISOLATOR
STAR/DELTA
LINKS REMOVED
FLEXIBLE CONDUIT
TEST
INSULATION TESTER
MΩ
>999
ZERO
AH2005-2

Fig: 15

TEST 4
PROTECTION BY SEPARATION OF CIRCUITS

Depicted in Fig: 16 on page 37 is an example of checking the separation of circuits.

All loads must be disconnected and switches turned on. Remember to remove any voltage sensitive electronic devices in the circuit such as dimmer switches, timers, controllers, electronic starters and controlgear for discharge lamps to prevent damage and misleading results.

The sequence is as follows:

- Remove all loads
- Ensure Isolation
- Select an Insulation continuity tester and set to 500V scale
- Place instrument on the circuit as shown
- Take reading and record at point FF on record of test results page 65

The reason for carrying out insulation resistance test is to prove that the circuits are electrical separate from each other. Tests should be made between conductors for all circuits.

This one test will consume a great deal of time. Take a standard domestic 6 way consumer unit. There are 6 circuits each having 3 conductors (we can ignore the return conductors in ring final circuits) this makes 18 separate conductors to check. (See page 70)

This table gives the minimum values allowed

CIRCUIT NOMINAL VOLTAGE (V)	TEST VOLTAGE d.c. (V)	MINIMUM INSULATION RESISTANCE (MΩ)
SELV AND PELV *	250	0.25
UP TO AND INCLUDING 500V WITH THE EXCEPTION OF THE ABOVE SUPPLIES	500	0.5
ABOVE 500V	1000	1.0

*SELV = Separated Extra-low Voltage PELV = Protected Extra-low Voltage

The IEE Guidance notes on testing state that if a reading of between that stated in the table and 2MΩ is obtained then further investigation is required.

Test: Separation of circuits
Reason: To check for damage to cables and a breakdown in insulation and interconnections between circuits.

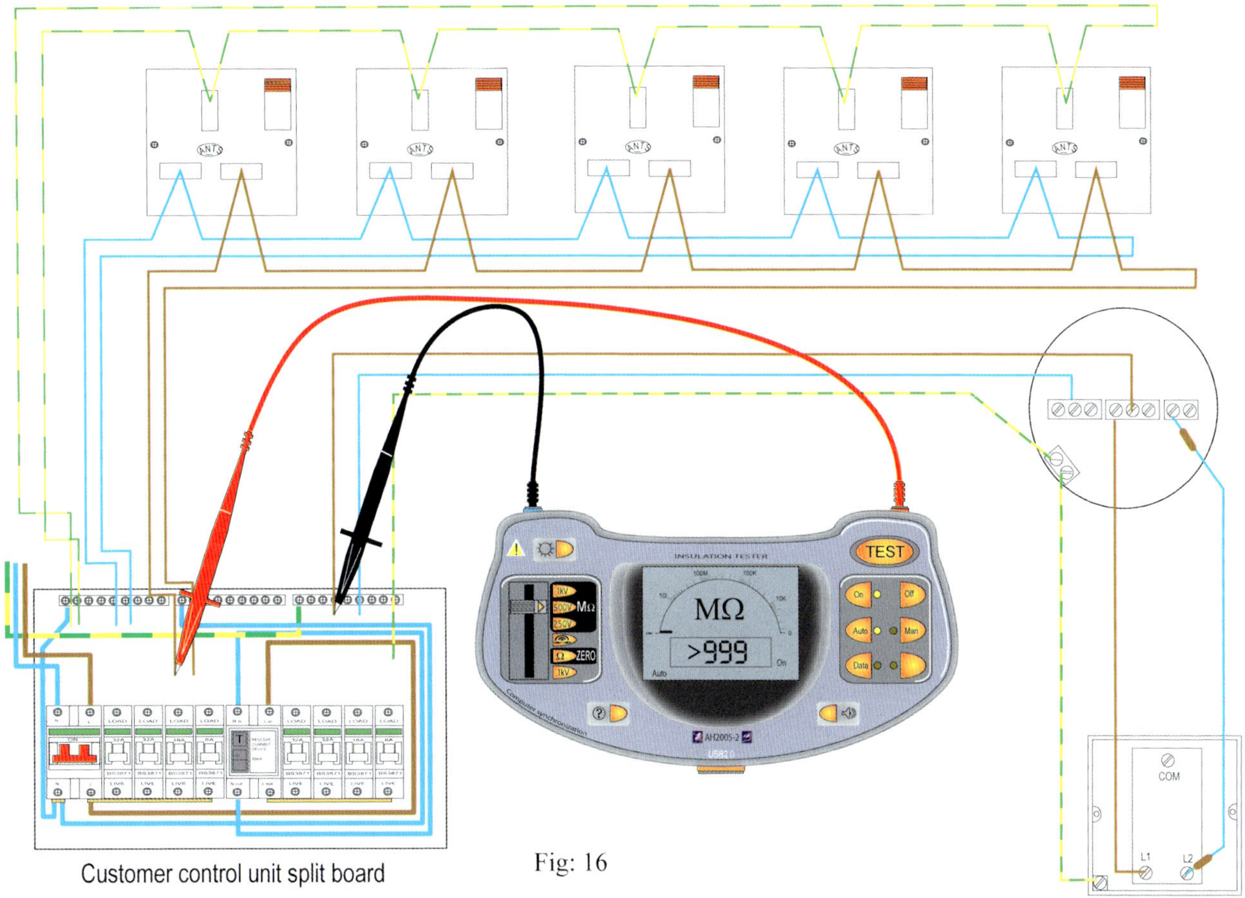

Fig: 16

TEST 5
POLARITY IN LIGHTING CIRCUITS

Depicted in Fig: 17 on page 39 is an example of checking the polarity of lighting circuits.

The sequence is as follows:

- Remove all loads
- Ensure Isolation
- Place links in circuit as indicated in Fig: 17
- Select an Insulation continuity tester and set to measure Ωs, zero instrument
- Place instrument on the circuit as shown
- Switch the switch on and off
- Take readings and record at point GG on record of test results page 65

Note: The reason for carrying out polarity is to prove that single pole devices and the centre pin on ES fittings are placed in the phase conductor.

Remember to remove all links before reinstating the supply.

Lighting circuit
Test: Polarity
Reason: To check that the centre pin on ES fittings and single pole devices are connected in the phase conductor and not in any other conductors in the circuit.

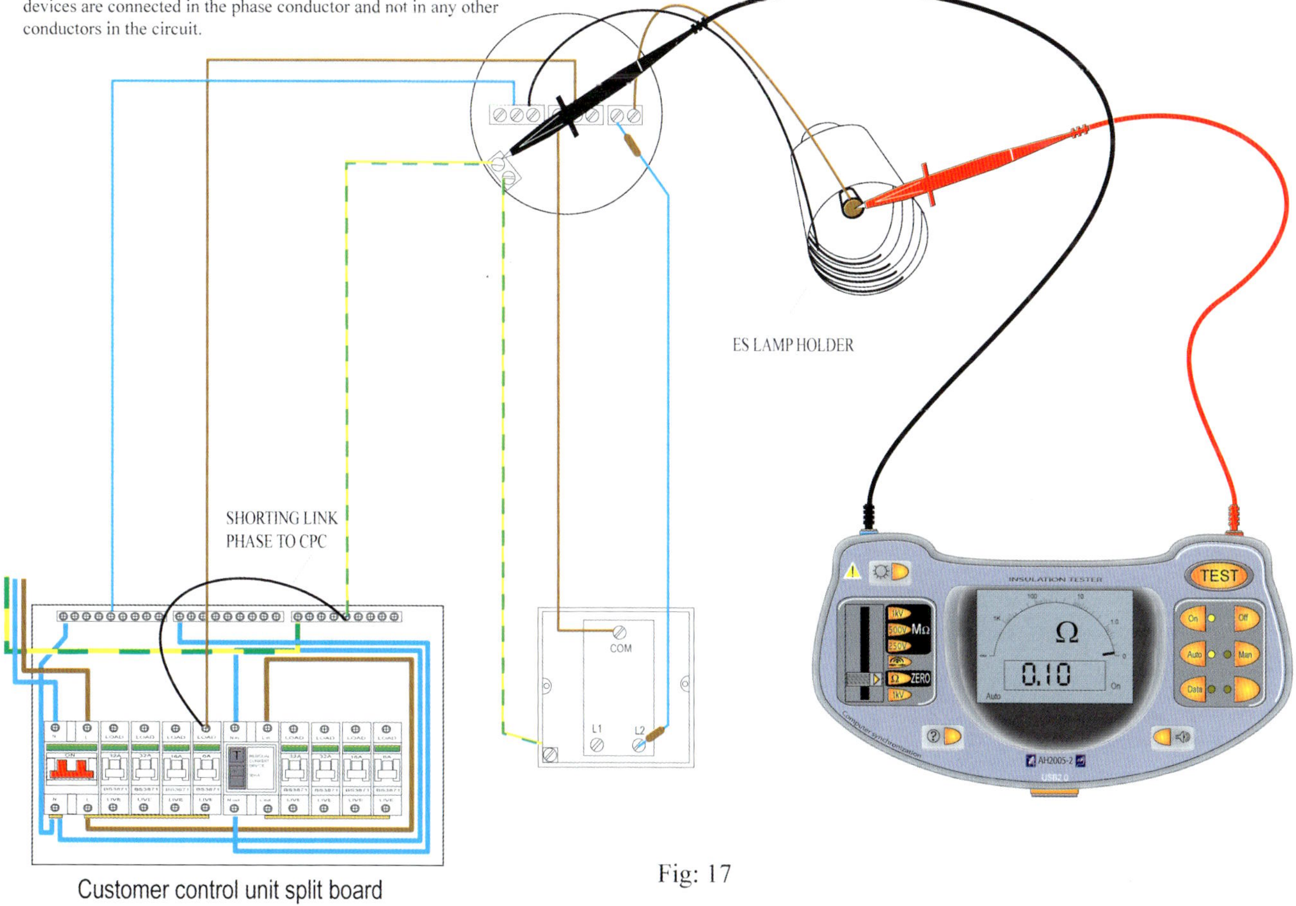

Fig: 17

TEST 6
EARTH FAULT LOOP IMPEDANCE IN LIGHTING CIRCUITS (Zs)

Depicted in Fig: 18 on page 41 is an example of checking the earth loop impedance in lighting circuits.

The sequence is as follows:

- Remove all loads
- Remove main equipotential bonding conductors (see note inside back cover)
- Ensure Isolation
- Select an earth loop tester and set to 20Ωs scale
- Place instrument in the circuit as shown
- Switch on the circuit
- Take readings and record at point II on record of test results page 65

This is classified as **live** testing so **precautions must be taken** to ensure your and everyone's safety during the tests.

You have previously carried out continuity of the cpc (R2) and established readings for the phase conductor (R1), we can now compare those readings with the readings established for earth loop impedance. Zs = Ze + (R1 + R2)

This test must be carried out at the remote point in the circuit, the remote point is the last ceiling outlet in the circuit using a special lead set conforming to HSE GS38.

If your instrument has various Ω scales you can start on the largest Ωs scale and work down the scales until an effective reading is obtained, but this need not be done if you have checked the regulations for maximum values for earth loop impedance in tables 41B2 and 41D. Remember that you need to know the disconnection time for the circuit; generally socket outlets should disconnect the circuit during earth faults within 0.4 of a second and fixed equipment within 5 seconds. In certain circumstances you can increase the disconnection time to 5 seconds for socket outlets, see Regulation 413-02-12 in BS 7671.

Once Zs and voltage have been established at the remote point in the circuit you can divide the voltage by Zs to give you Ia (fault to earth current). Apply this calculation to the time current characteristic graphs in the regulations and you will be able to determine the actual disconnection time for the circuit.

Lighting circuit
Test: Earth loop impedance
Reason: Check to make sure that the circuit meets the requirements for Zs as stated in BS7671. Zs = Ze + (R1 + R2)
All loads removed and circuit made live

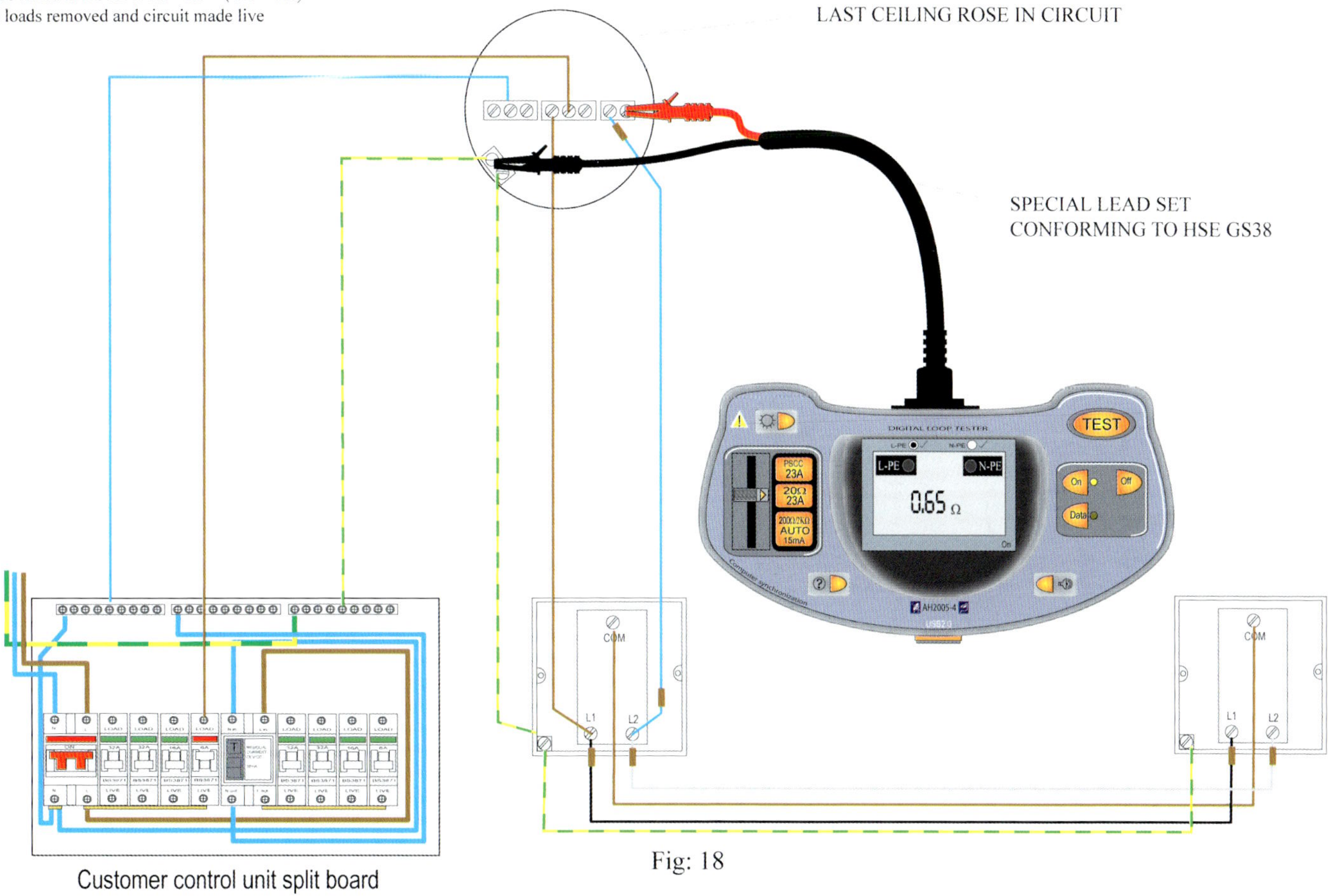

Fig: 18

EARTH FAULT LOOP IMPEDANCE IN RING FINAL CIRCUITS (Zs)

Depicted in Fig: 19 on page 43 is an example of checking the earth loop impedance in ring circuits.

The sequence is as follows:

- Remove all loads
- Remove main equipotential bonding conductors (see note inside back cover)
- Ensure Isolation
- Select an earth loop tester and set to largest Ωs scale
- Place instrument in the circuit as shown
- Switch on the circuit
- Take readings and record at point II on record of test results page 65

This is classified as **live** testing so **precautions must be taken** to ensure your and everyone's safety during the tests.

You have previously carried out continuity of the cpc (R2) and established readings for the phase conductor (R1), we can now compare those readings with the readings established for earth loop impedance. Zs = Ze + (R1 + R2)

This test must be carried out at the remote point in the circuit, The remote point in a ring final circuit is the centre of the ring. You can confirm this by taking various readings and recording the highest reading on the test documentation.

If your instrument has various Ω scales you can start on the largest Ωs scale, and work down the scales until an effective reading is obtained, but this need not be done if you have checked the regulations for maximum values for earth loop impedance in tables 41B1 and 41B2. Remember that you need to know the disconnection time for the circuit; generally socket outlets should disconnect the circuit during earth faults within 0.4 of a second and fixed equipment within 5 seconds.

Once Zs and voltage have been established at the remote point in the circuit you can divide the voltage by Zs which will give you Ia (fault to earth current). Apply this calculation to the time current characteristic graphs in the regulations and you will be able to determine the actual disconnection time for the circuit.

Ring Final Circuits

Test: Earth loop impedance

Reason: Check to make sure that the circuit meets the requirements for Zs as stated in BS7671. Zs = Ze + (R1 + R2)

All loads removed and circuit made live

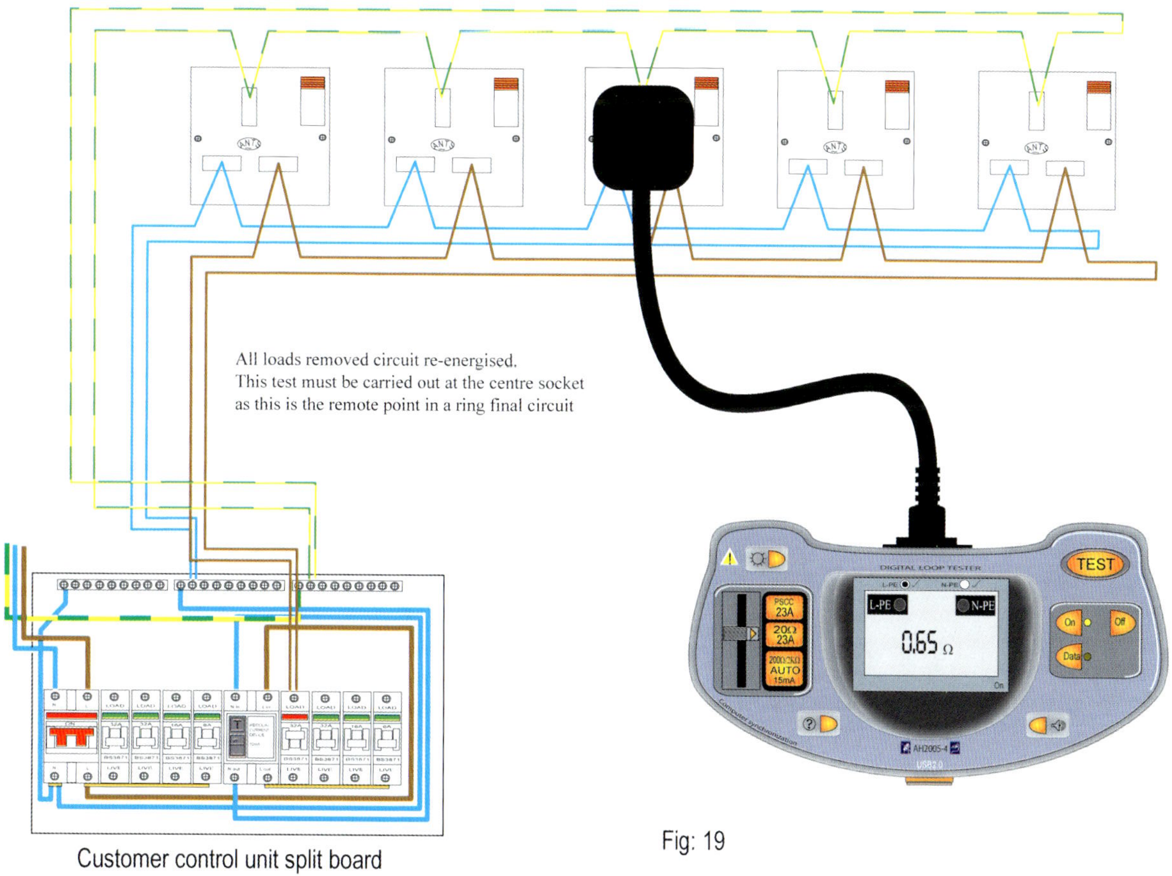

Fig: 19

EARTH FAULT LOOP IMPEDANCE IN 3 PHASE MOTOR CIRCUITS (Zs)

Depicted in Fig: 20 on page 45 is an example of checking the earth loop impedance in 3 phase motor circuits.

The sequence is as follows:

- Remove all loads
- Remove main equipotential bonding conductors(see note inside back cover)
- Ensure Isolation
- Select an earth loop tester and set to largest Ωs scale
- Place instrument on the circuit as shown
- Switch on the circuit
- Take readings and record at point II on record of test results page 65

This is classified as **live** testing so **precautions must be taken** to ensure your and everyone's safety during the tests. To carry out the test you can either remove the star/delta links or disconnect the conductors. Bear in mind if you disconnect the conductors make sure that you insulate the ones that are free. **Only carry out the test on one phase at a time or you could seriously damage your meter**

You will need a special lead set conforming to HSE GS38. Some instruments have test leads with earth and neutral clips on the test lead these should be clamped on the casing of the motor, the neutral is only there to power the meter and give a voltage reading. Make sure you carry out the test on all three phases and the results should be substantially the same. Having previously carried out continuity of the cpc (R2) and established readings for the phase conductor (R1), we can now compare those readings with the readings established for earth loop impedance.Zs = Ze + (R1 + R2) This test must be carried out at the motor as this is the remote point in the circuit.

If your instrument has various Ω scales you can start on the largest Ωs scale, and work down the scales until an effective reading is obtained, but this need not be done if you have checked the regulations for maximum values for earth loop impedance in tables 41B2 and 41D. Remember that you need to know the disconnection time for the circuit, generally socket outlets should disconnect the circuit during earth faults within 0.4 of a second and fixed equipment within 5 seconds.
Once Zs and voltage have been established at the remote point in the circuit you can divide the voltage by Zs to give you Ia (fault to earth current). Apply this calculation to the time current characteristic graphs in the regulations and you will be able to determine the actual disconnection time for the circuit.

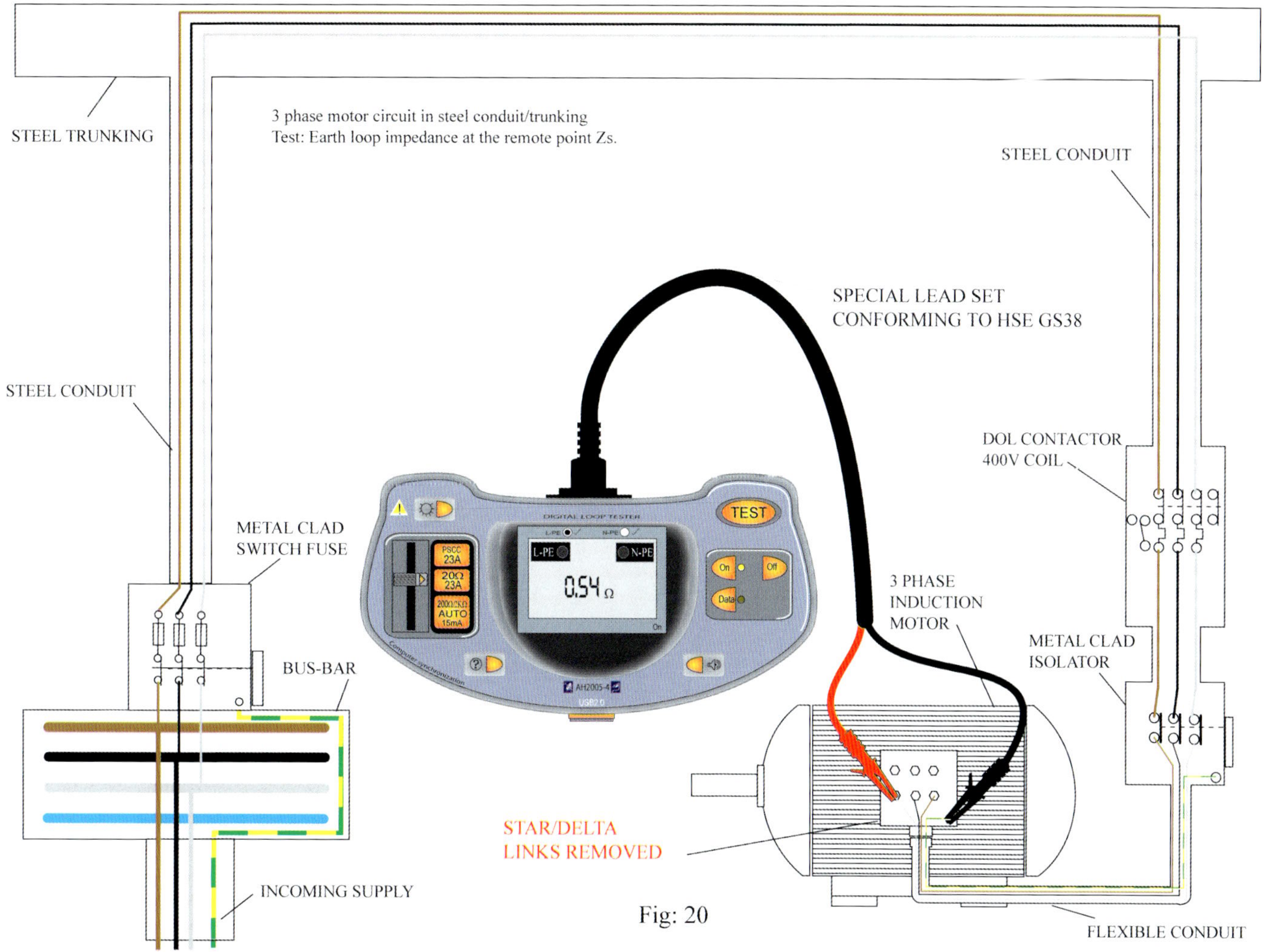
STEEL TRUNKING
3 phase motor circuit in steel conduit/trunking
Test: Earth loop impedance at the remote point Zs.
STEEL CONDUIT
SPECIAL LEAD SET
CONFORMING TO HSE GS38
STEEL CONDUIT
DOL CONTACTOR
400V COIL
METAL CLAD
SWITCH FUSE
DIGITAL LOOP TESTER
TEST
L-PE
N-PE
0.54 Ω
PSCC 23A
20Ω 23A
2000Ω/2KΩ AUTO 15mA
On
Off
Data
3 PHASE
INDUCTION
MOTOR
METAL CLAD
ISOLATOR
BUS-BAR
AH2005-4
USB2.0
Computer synchronization
STAR/DELTA
LINKS REMOVED
INCOMING SUPPLY
FLEXIBLE CONDUIT

Fig: 20

MEASUREMENT OF EXTERNAL EARTH FAULT LOOP IMPEDANCE (Ze)

Depicted in Fig: 21 on page 47 is an example of checking the external earth loop impedance.

The sequence is as follows:

- Remove all loads
- Ensure Isolation
- Disconnect main earth conductor (prevents parallel paths)
- Select an earth loop tester and set to largest Ωs scale
- Place instrument on the circuit as shown
- Switch on the circuit
- Take readings and record at point D on record of test results page 64

This is classified as **live** testing so **precautions must be taken** to ensure your and everyone's safety during the tests. To carry out the test you can place the test leads on the supply point or incoming supply. Make sure that you are going to connect into the supply after the main switch, this will enable isolation of the conductors thus making the test safer than if you connected on to the main tails or supply conductors. Three phase can be carried out in the same manner, but you must **only carry out the test on one phase at a time or you could seriously damage your meter**

You will need a special lead set conforming to HSE GS38,

This test will give you a definitive result for the external impedance thus enabling you to compare that quoted or calculated results are correct.

You have measured Zs and (R1+R2), so comparisons can now be made to establish the effectiveness of your previous test results.

Important note: Take great care with this test because the instrument is manufactured with an overload rating of up to 600V, and at this point in the circuit the transients can be greater than 600V.

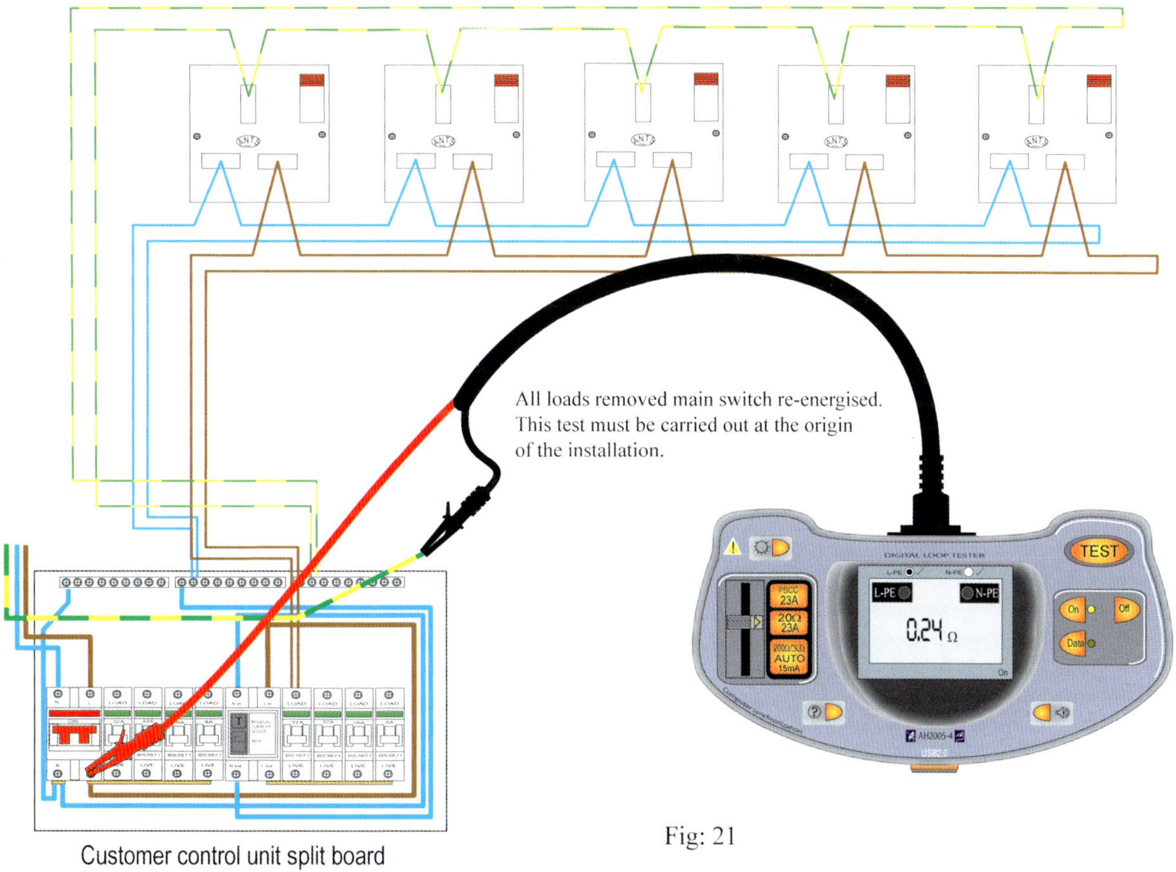
All loads removed main switch re-energised.
This test must be carried out at the origin
of the installation.
Customer control unit split board
TEST
On
Off
Data
DIGITAL LOOP TESTER
L-PE
N-PE
0.24 Ω
PSCC 23A
20Ω 23A
AUTO 15mA
AH2005-4
Computer synchronization

Fig: 21

TEST 7
FUNCTIONAL TESTING IN LIGHTING CIRCUITS
(RCDs)

Depicted in Fig: 22 on page 49 is an example of checking the earth leakage device in a lighting circuit.

The sequence is as follows:

- Remove all loads
- Ensure Isolation
- Select an RCD tester and set to tripping current of device
- Place instrument in the circuit as shown
- Switch on the circuit
- Take readings and record at points JJ, KK, LL, MM, NN, OO, PP on the record of test results page 65

This is classified as **live** testing so **precautions must be taken** to ensure your and everyone's safety during the tests.

You will need a special lead set conforming to HSE GS38.

This test must be carried out at the remote point in the circuit in this case the last ceiling outlet.

In Fig: 22 the lighting circuit is protected by an RCD, so the RCD must be tested to prove that it disconnects the circuit within the time stated in the on-site guide.

Example

Set instrument to tripping current of RCD (stated on the front of the device). Set the multiplier on the instrument to 1 X, check polarity on the instrument and then press test button, the RCD should disconnect within 200ms. Change the phase angle from 0° to 180° and carry out the test again. Now set instrument to 1/2 X tripping current and carry out the test again, the trip should not trip This checks the trip for nuisance tripping. You should have carried out 4 tests so far. The final test is to operate the test button on the trip, the trip should operate. If any of the tests fail the trip should be changed for a new one. (See note: 6 page 78)

Assuming that on the 1 X trip you established a reading of 80ms at 0° and 65ms at 180°, the reading that you record for this test should be the higher of the two, which is 80ms.

Lighting circuit
Test: Residual current device diconnection times
Reason: Check to make sure that the RCD trips
within the time stated in the On-Site guide to BS7671
All loads removed and circuit supply re-instated

Test: IΔN on 0°& 180°
½ IΔN on 0°& 180°
150mA on 0°& 180°
Press trip test button on device

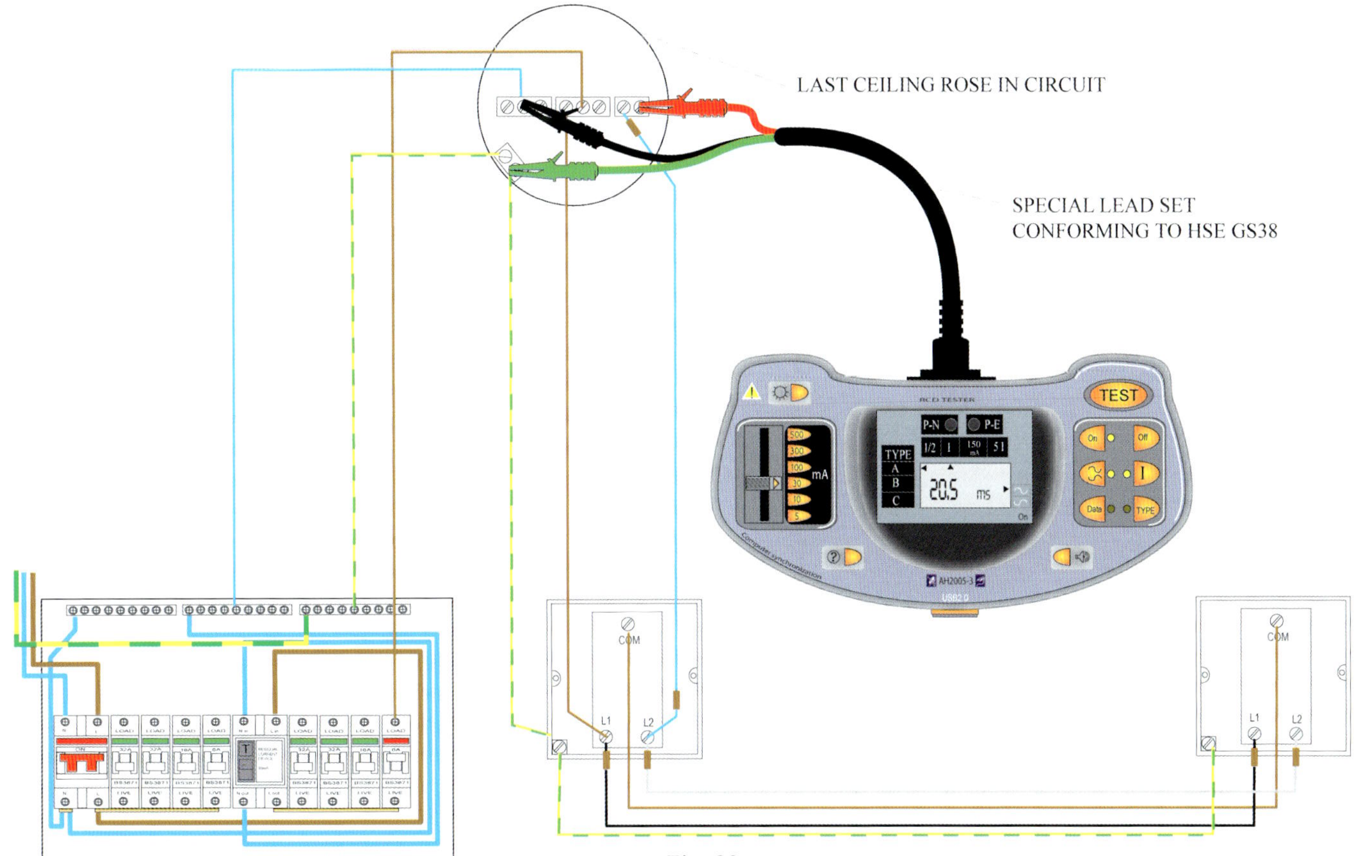

Fig: 22

FUNCTIONAL TESTING IN RING FINAL CIRCUITS (RCDs)

Depicted in Fig: 23 on page 51 is an example of checking the earth leakage device in a ring final circuit.

The sequence is as follows:

- Remove all loads
- Ensure Isolation
- Select an RCD tester and set to tripping current of device
- Place instrument in the circuit as shown
- Switch on the circuit
- Take readings and record at points JJ, KK, LL, MM, NN, OO, PP on the record of test results page 65

This is classified as **live** testing so **precautions must be taken** to ensure your and everyone's safety during the tests.

You can use the standard lead set supplied with the instrument.

This test must be carried out at the remote point in the circuit, refer to the method used when testing Earth Loop in ring circuits on page 42.

In Fig: 23 the ring circuit is protected by an RCD, so the RCD must be tested to prove that it disconnects the circuit within the time stated in the on-site guide.

Example

Set instrument to tripping current of RCD (stated on the front of the device). Set the multiplier on the instrument to 1 X, check polarity on the instrument and then press test button; the RCD should disconnect within 200ms. Change the phase angle from 0° to 180° and carry out the test again. Now set instrument to 1/2 X tripping current and carry out the test again, the trip should not trip This checks the trip for nuisance tripping. You should have carried out 4 tests so far. The final test is to operate the test button on the trip, the trip should operate. If any of the tests fail the trip should be changed for a new one. (See note: 6 page 78)

Assuming that on the 1 X trip you had a reading of 80ms at 0° and 65ms at 180°, the reading that you write down for this test should be the higher of the two, which is 80ms.

Ring final circuit
Test: Residual current device diconnection times
Reason: Check to make sure that the RCD trips
within the time stated in the On-Site guide to BS7671
All loads removed and circuit supply re-instated

Test: IΔN on 0°& 180°
$\frac{1}{2}$IΔN on 0°& 180°
150mA on 0°& 180°
Press trip test button on device

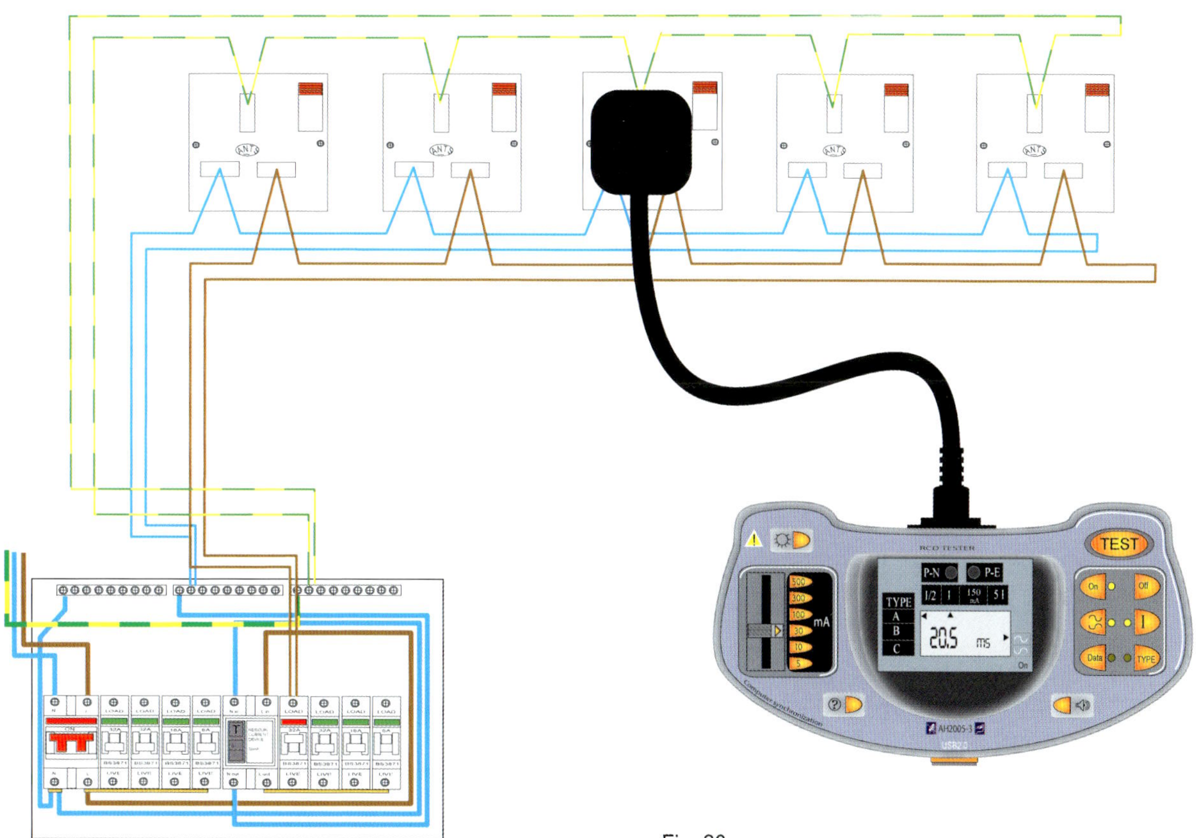

Customer control unit split board

Fig: 23

MEASUREMENT OF THE PROSPECTIVE SHORT CIRCUIT CURRENT (P-N) AT THE ORIGIN OF THE SUPPLY

Depicted in Fig: 24 on page 53 is an example of measuring the prospective short circuit current at the origin of the supply. **See important note on page 46**

The sequence is as follows:

- Remove all loads
- Ensure Isolation
- Select an earth loop tester, set to PSCC or 20Ω scale (see note 8 page 78)
- Place instrument on the circuit as shown
- Switch on the circuit
- Take readings and record

Note: 1. This is classified as **live** testing so **precautions must be taken** to ensure your and everyone's safety during the tests. To carry out the test place the test leads on the supply point or incoming supply. Make sure that you connect into the supply after the main switch, this will enable isolation of the conductors and make the test safer. Three phase can be carried out in the same manner, but you must **only carry out the test on one phase at a time or you could seriously damage your meter**

If your instrument is capable of measuring PSCC then calculations will not be necessary just select the PSCC scale as indicated and test. If the lead set has phase, neutral and earth connections then the neutral and earth leads need to be attached to the neutral terminal.

This test will give you a calculated figure for the prospective short circuit current. The reason for carrying out the test is to check that in the event of a short circuit the current is of a lesser value than the rated PSCC of the switchgear.

Example using calculations from resistance measurement

Assume that you have a reading of 0.01Ω, you then need to divide the voltage at this point by the resistance, e.g. 230 ÷ 0.01Ω = 23kA. You then need to check the rating of the main switchgear (written on manufacturers label). Assume that the swithgear has a PSCC value of 80kA, then under short circuit conditions the switchgear will sustain the fault without damage. If the rating of the switchgear was 10kA then it is likely that in the event of a short circuit the switchgear would disintegrate possibly causing fire.

Test: Prospective short circuit current at the origin phase - neutral

Using your loop tester select the PSCC range test and record result

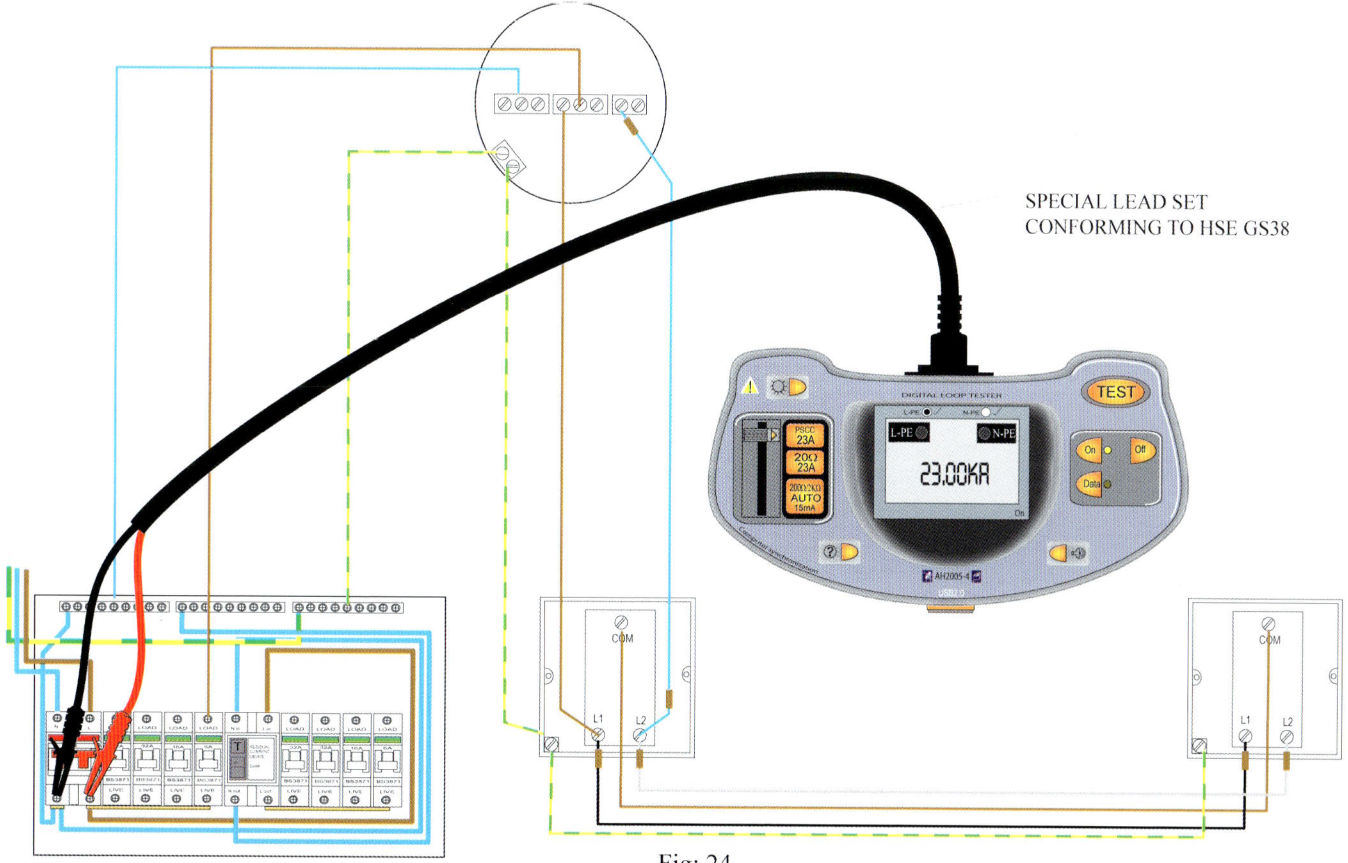

Customer control unit split board

Fig: 24

<u>MEASUREMENT OF THE PROSPECTIVE SHORT CIRCUIT CURRENT (P-N) AT THE ORIGIN OF A CIRCUIT</u>

Depicted in Fig: 25 on page 55 is an example of measuring the prospective short circuit current at the origin of the circuit. **See important note on page 46**

The sequence is as follows:

- Remove all loads
- Ensure Isolation
- Select an earth loop tester, set to PSCC or 20Ω scale (see note 8 page 78)
- Place instrument on the circuit as shown
- Switch on the circuit
- Take readings and record at point SS on record of test results page 65

This is classified as **live** testing so **precautions must be taken** to ensure yours and everyone's safety during the tests. To carry out the test place the test leads on supply after the protective device feeding the circuit you are testing, this will enable disconnection of the phase conductor. Three phase can be carried out in the same manner, but you must **<u>only carry out the test on one phase at a time or you could seriously damage your meter</u>**

If your instrument is capable of measuring PSCC then calculations will not be necessary just select the PSCC scale as indicated and test. If the lead set has phase, neutral and earth connections then the neutral and earth leads need to be attached to the neutral terminal.

This test will give you a calculated figure for the prospective short circuit current. The reason for carrying out the test is to check that in the event of a short circuit the current is of a lesser value than the rated current of the protective device. Note: BS88 fuses are generally 80kA, MCB's between 1 and 25kA

<u>Example using calculations from resistance measurement</u>

Assume that you have a reading of 0.28Ω, you then need to divide the voltage at this point by the resistance, e.g. 230 ÷ 0.28Ω = 821A. You then need to check the rating of the main switchgear (written on manufacturers label). Assume that the swithgear has a PSCC value of 1kA, then under short circuit conditions the device will sustain the fault without damage. If the rating of the device was .5kA then it is likely that in the event of a short circuit the device would disintegrate possibly causing fire.

Test: Prospective short circuit current at the origin phase - neutral

Using your loop tester select the PSCC range test and record result.
Reason: Check to make sure that the short circuit current does not exceed the breaking capacity of the protective device

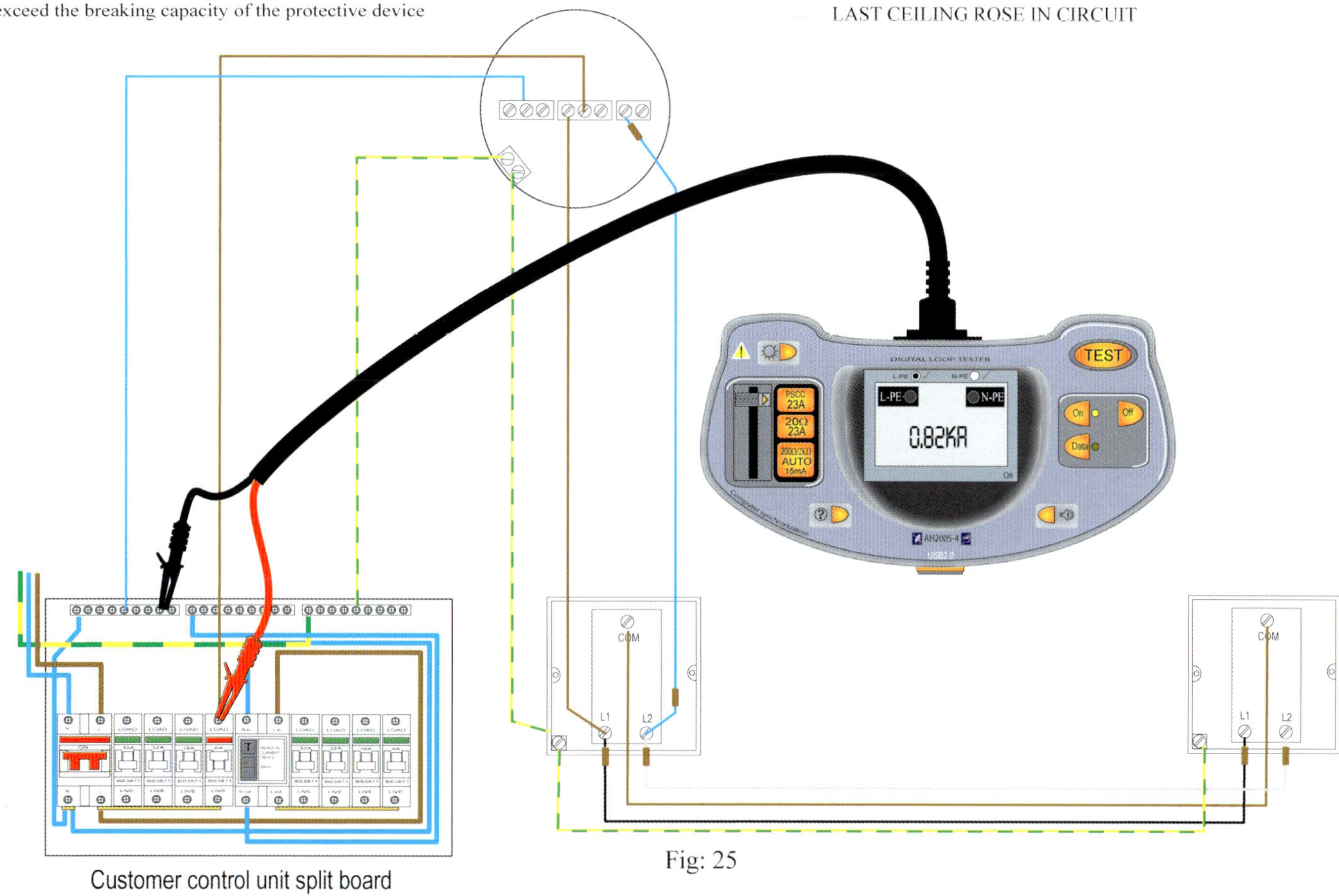

Fig: 25

MEASUREMENT OF THE PROSPECTIVE EARTH FAULT CURRENT IN LIGHTING CIRCUITS AT THE REMOTE POINT (Ia)

Depicted in Fig: 26 on page 57 is an example of measuring the prospective earth fault current at the remote point in the circuit.

The sequence is as follows:

- Remove all loads
- Ensure Isolation
- Select an earth loop tester, set to PSCC or 20Ω scale (see note 8 page 78)
- Place instrument on the circuit as shown
- Switch on the circuit
- Take readings and record at point RR on record of test result page 65

This is classified as **live** testing so **precautions must be taken** to ensure your and everyone's safety during the tests. To carry out the test place the test leads on the conductors at the remote point, it may be necessary to remove the ceiling rose.

If your instrument is capable of measuring PEFC then calculations will not be necessary just select the PSCC scale and test. If the lead set has phase, neutral and earth connections then the neutral and earth leads need to be attached to the neutral and earth terminal respectively.

This test will give you a calculated figure for the prospective earth fault current. The reason for carrying out the test is to check that in the event of an earth fault the current is of a sufficiently high enough value to operate the protective device within the time stated in the Regulations.

Example using calculations from resistance measurement

Assume that you have a reading of 0.96Ω, you then need to divide the voltage at this point by the resistance, e.g. 230 ÷ 0.96Ω = 239A. You then need to check the BS number and type of the protective device (written on the device). Assume that the protective device is a 6A BS3871 type 2, with the reading of 239A, you can now cross check the time current graphs in the Regulations to get an actual disconnection time for the circuit. In this case it is less than 0.01 of a second. As the circuit we have tested is a lighting circuit, which has a maximum disconnection time of 5 seconds, then this circuit disconnects within that time and thus passes the test.

Test: Prospective earth fault current at the remote point phase - earth

Using your loop tester select the PSCC range test and record result.
Reason: Check to make sure that the earth fault current is sufficiently high enough value to disconnect the protective device within the time stated in BS7671

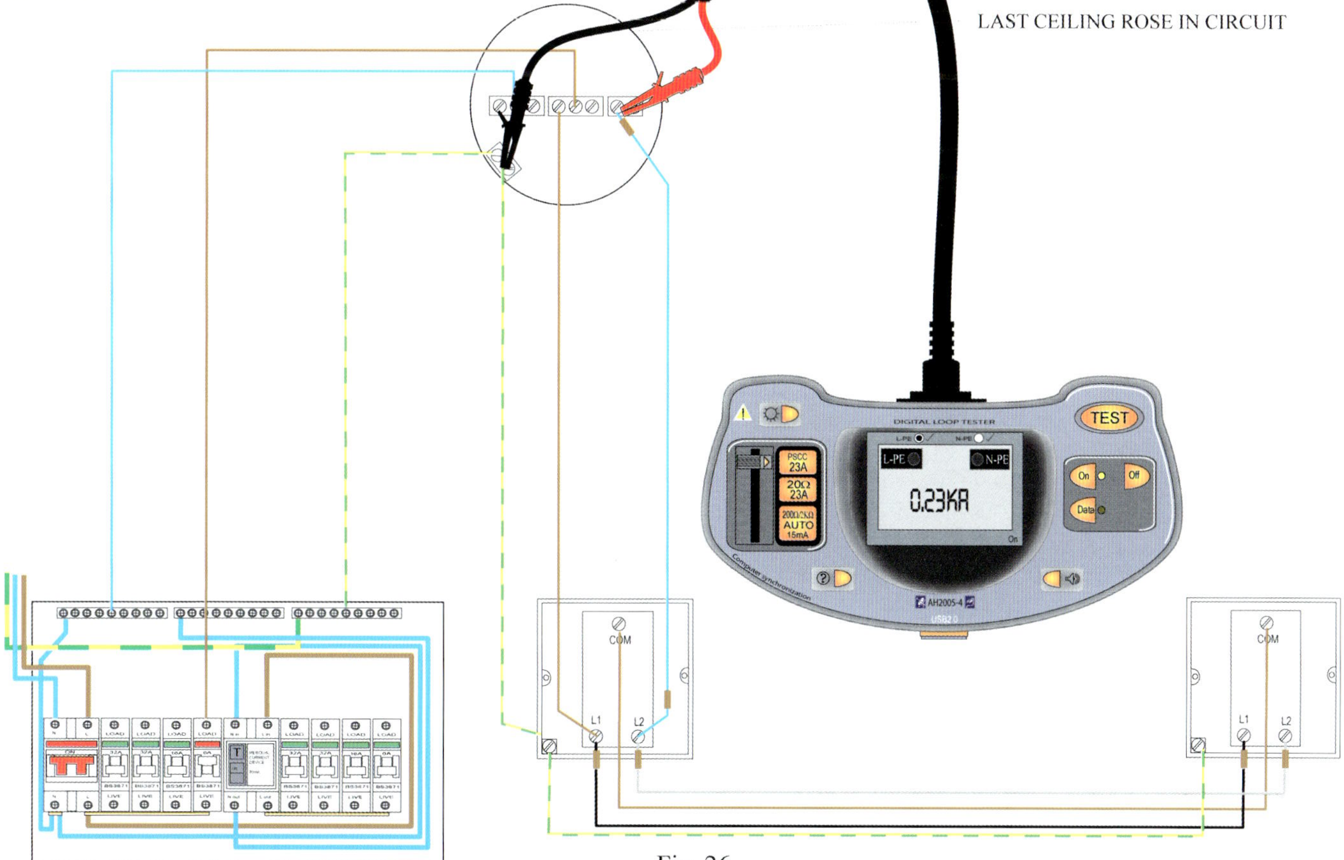

Fig: 26

MEASUREMENT OF THE PROSPECTIVE EARTH FAULT CURRENT IN RING CIRCUITS AT THE REMOTE POINT (Ia)

Depicted in Fig: 27 on page 59 is an example of measuring the prospective earth fault current at the remote point in the circuit.

The sequence is as follows:

- Remove all loads
- Ensure Isolation
- Select an earth loop tester, set to PSCC or 20Ω scale (see note 8 page 78)
- Place instrument on the circuit as shown
- Switch on the circuit
- Take readings and record at point RR on record of test result page 65

This is classified as **live** testing so **precautions must be taken** to ensure your and everyone's safety during the tests. To carry out the test plug the test lead in to the centre socket on the ring final circuit. You will have already found the centre socket when carrying out test 6 page 42.

If your instrument is capable of measuring PEFC then calculations will not be necessary just select the PSCC scale and test. If the lead set has phase, neutral and earth connections then the neutral and earth leads need to be attached to the neutral and earth terminal respectively.

This test will give you a calculated figure for the prospective earth fault current. The reason for carrying out the test is to check that in the event of a earth fault the current is of a sufficiently high enough value to operate the protective device within the time stated in the Regulations.

Example using calculations from resistance measurement

Assume that you have a reading of 0.64Ω, you then need to divide the voltage at this point by the resistance, e.g. 230 ÷ 0.64Ω = 359A. You then need to check the BS number and type of the protective device (written on the device). Assume that the protective device is a 32A BS3871 type 2, with the reading of 359A, you can now cross check the time current graphs in the Regulations to get an actual disconnection time for the circuit. In this case it is less than 0.03 of a second. As the circuit we have tested is a ring final circuit, which has a maximum disconnection time of 0.4 second, then this circuit disconnects within that time and thus passes the test.

Ring final circuits
Test: Prospective earth fault current phase to earth at the remote point
Reason: Check to make sure that the fault current to earth is sufficiently high enough value to disconnect the protective device within the time stated in BS7671

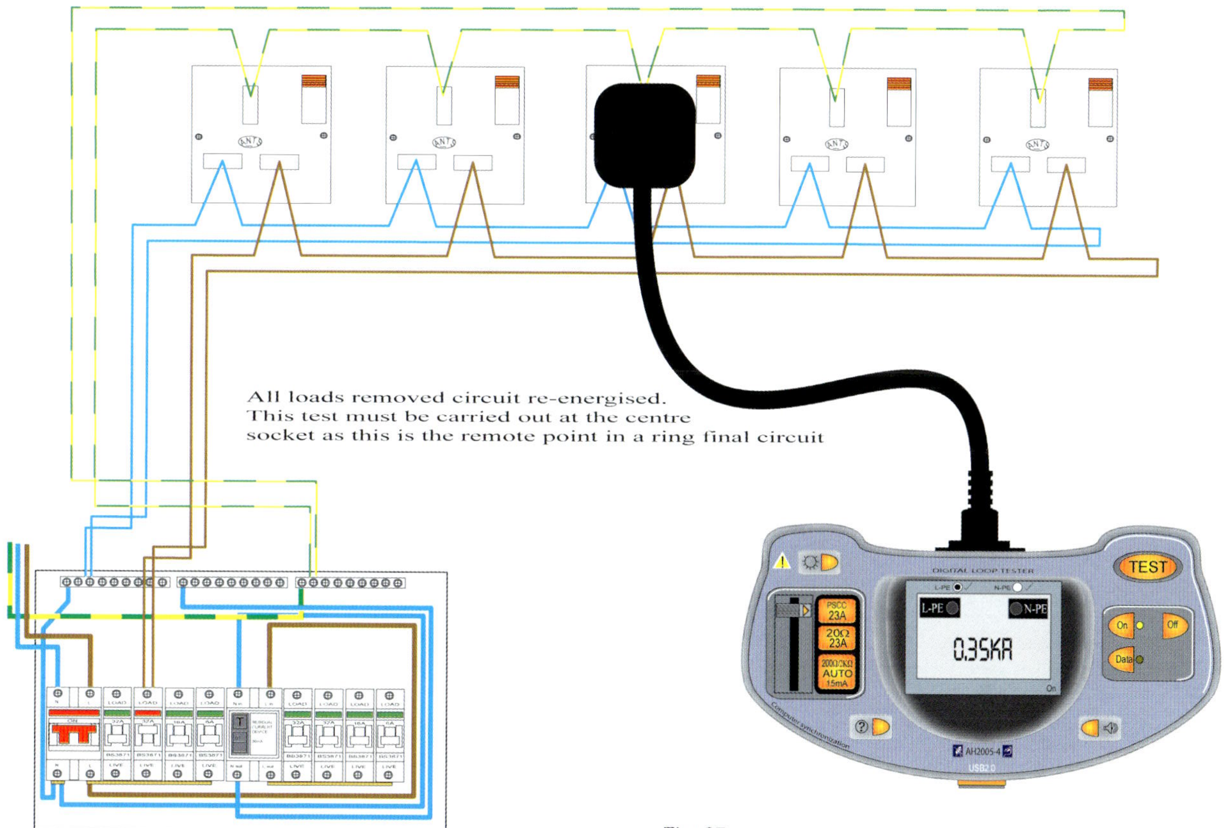

Customer control unit split board

Fig: 27

MEASUREMENT OF THE PROSPECTIVE EARTH FAULT CURRENT IN A 3 PHASE MOTOR CIRCUIT AT THE REMOTE POINT (Ia)

Depicted in Fig: 28 on page 61 is an example of measuring the prospective earth fault current at the remote point in the circuit.

The sequence is as follows:

- Remove all loads
- Ensure Isolation
- Select an earth loop tester, set to PSCC or 20Ω (see note 8 page 78)
- Place instrument on the circuit as shown
- Switch on the circuit
- Take readings and record at point RR on record of test result page 65

This is classified as **live** testing so **precautions must be taken** to ensure your and everyone's safety during the tests. To carry out the test you must place the test leads on the conductors at the remote point, in this case that is at the motor.

If your instrument is capable of measuring PEFC then calculations will not be necessary just select the PSCC scale and test. If the lead set has phase, neutral and earth connections then the neutral and earth leads need to be attached to the earth terminal on the motor.

This test will give you a calculated figure for the prospective earth fault current. The reason for carrying out the test is to check that in the event of an earth fault the current is of a sufficiently high enough value to operate the protective device within the time stated in the Regulations.

Example using calculations from resistance measurement

Assume that you have a reading of 0.54Ω, you then need to divide the voltage at this point by the resistance, e.g. 230 ÷ 0.54Ω = 426A. You then need to check the BS number and type of the protective device (written on the device). Assume that the protective device is a 16A BS3871 type 2, with the reading of 426A you can now cross check the time current graphs in the Regulations to get an actual disconnection time for the circuit, in this case it is less than 0.1 of a second.
As the circuit you have tested is a motor circuit, which has a maximum disconnection time of 5 seconds, then this circuit disconnects within that time and thus passes the test. **Only carry out the test on one phase at a time or you could seriously damage your meter.**

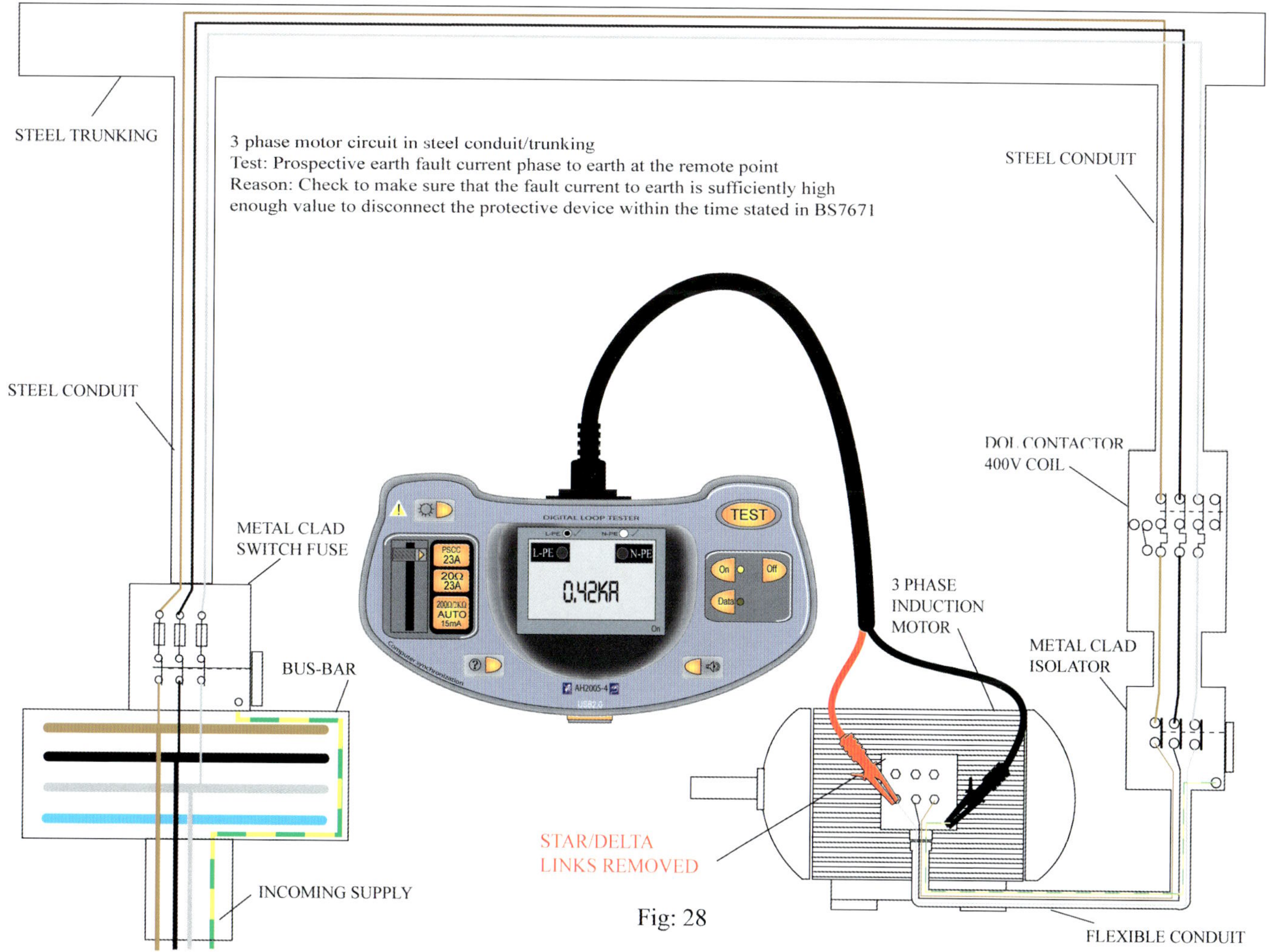
3 phase motor circuit in steel conduit/trunking
Test: Prospective earth fault current phase to earth at the remote point
Reason: Check to make sure that the fault current to earth is sufficiently high enough value to disconnect the protective device within the time stated in BS7671
STEEL TRUNKING
STEEL CONDUIT
STEEL CONDUIT
DOL CONTACTOR
400V COIL
METAL CLAD
SWITCH FUSE
BUS-BAR
INCOMING SUPPLY
3 PHASE
INDUCTION
MOTOR
METAL CLAD
ISOLATOR
STAR/DELTA
LINKS REMOVED
FLEXIBLE CONDUIT
DIGITAL LOOP TESTER
TEST
L-PE
N-PE
0.42KA
PSCC 23A
20Ω 23A
2000/2KΩ AUTO 15mA
On
Off
Data
AH2005-4

Fig: 28

HOW DO RCD's WORK?

Depicted in Fig: 29 on page 63 is the internal construction of an RCD (Residual Current Device).

An RCD is basically made up of 3 coils wrapped around a soft iron core. The two larger coils, Phase and Neutral, carry the load currents for the circuit or circuits that it is protecting, the third coil continually monitors the soft iron core checking to see if any flux is circulating in the core. Basically operating like a transformer.

The current I1 flowing in the phase conductor must be the same as I2 returning in the neutral conductor, if there is a difference in the currents I1 and I2 then a flux will circulate in the iron core. The third coil continually monitors the soft iron core checking to see if any flux is circulating. If a flux is present then this induces a voltage in the third coil which in turn is then amplified and the trip coil breaks the double pole switch.

The difference between I1 and I2 must be equal to or greater than the actual tripping current of the device. However, this actual tripping current is often much less than the rated tripping current, e.g. many RCDs with a rated tripping current of 30mA will actually trip at currents as low as 22mA.

The test button on the device is connected from the incoming phase through a resistor to the outgoing neutral. When the button is depressed the resistor allows a current to by-pass the phase coil thus causing an imbalance in the core in turn causing the trip to operate.

Common occurrence

The trip operates and when re-instated is all right.

The reason for this is that there may be an accumulation of leakage currents to earth from domestic appliances. For instance, if you had a cooker with 10mA leakage, kettle with 10mA leakage and an immersion heater with 10mA leakage, if all of these where to be on at the same time you have accumulated leakage of 30mA which will operate the trip. Cooker elements and immersion heaters are the most common causes of nuisance tripping.

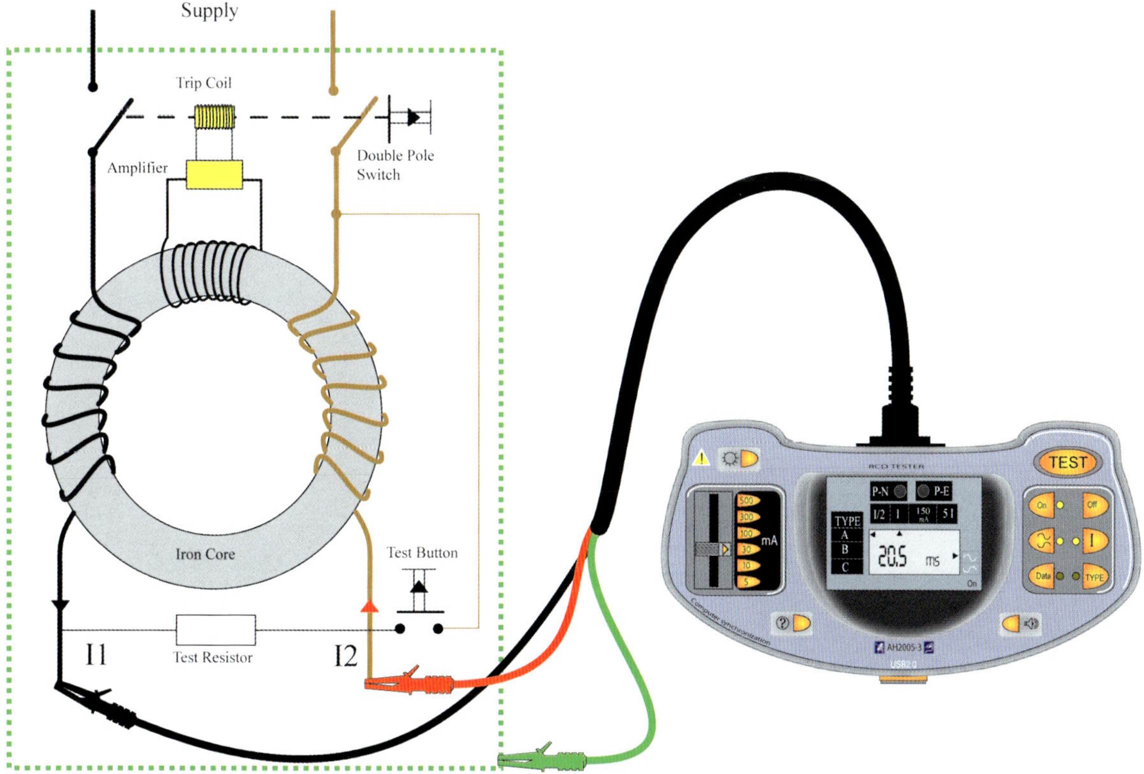
Supply
Trip Coil
Amplifier
Double Pole
Switch
Iron Core
Test Button
I1
Test Resistor
I2
RCD TESTER
TEST
P-N
P-E
I/2
I
150
mA
5 I
TYPE
A
B
C
20.5
ms
On
Off
Data
TYPE
500
300
100
30
10
5
mA
AH2005-3
USB2.0

Fig: 29

RECORD OF TEST RESULTS AND DESIGN DETAILS

Testers surname	A		Instruments used	
Testers first name	A		Make	Seriel No
Date of tests	B	Continuity	F	G
Earthing Arrangements	C	Insulation	F	G
External impedance (Ze)	D	Loop impedance	F	G
Supply voltage	E	RCD	F	G

Type of premises	Domestic		Commercial		Industrial	
Estimated age of installation	Years		Date last inspected			
Evidence of alterations	Yes		No		Not apparent	

Circuit referance			H										
Number of points served													
Protective Device	Device BS		J										
	Type		K										
	Rating (A)		L										
Phase (colour or 3 phase)			M										
Cable size	Phase (mm²)		N										
	CPC (mm²)		O										
Continuity	Bonding (Ω)		P										
	CPC (Ω)		Q										
	Radial	Design R1 + R2 (Ω)	R										
		Measured R1 + R2 (Ω)	S										
	Ring	Design R1 + R2 (Ω)	T										
		Measured R1 + R2 (Ω)	U										
Confirmation of ring Y or N			V										

RECORD OF TEST RESULTS AND DESIGN DETAILS

Insulation resistance in MΩ	Phase to CPC	W											
	Phase to neutral	X											
	Neutral to CPC	Y											
	L1 to CPC	Z											
	L2 to CPC	AA											
	L3 to CPC	BB											
	L1 to neutral	CC											
	L2 to neutral	DD											
	L3 to neutral	EE											
Separation of circuits			FF										
Polarity (Pass or Fail)		GG											
Earth fault loop Zs (Ω)	Maximum design Zs (Ω)	HH											
	Actual Measured Zs (Ω)	II											
RCD	Rated IΔN (mA)	JJ											
	$\frac{1}{2}$ ΔN Y/N	KK											
Trip time (ms) at	IΔN trip ms 0°	LL											
	IΔN trip ms 180°	MM											
	150mA ms 0°	NN											
	150mA ms 180°	OO											
	Test button Y/N	PP											
PEFC (Ia)	Origin	QQ											
	Remote	RR											
PSCC (P-N)	Origin	SS											

RECORD OF TEST RESULTS AND DESIGN DETAILS

Testers surname	Hinsley			Instruments used		
Testers first name	Tony			Make	Seriel No	
Date of tests	16 May 2006	Continuity		AH2005-1	1234567-4	
Earthing Arrangements	TN-C-S	Insulation		AH2005-2	356729-2	
External impedance (Ze)	0.19Ω	Loop impedance		AH2005-3	345552779	
Supply voltage	230v	RCD		AH2005-4	3569874-2	
Type of premises	Domestic	✔	Commercial		Industrial	
Estimated age of installation	Years	15	Date last inspected	Not known		
Evidence of alterations	Yes		No	✔	Not apparent	

Circuit referance			Cooker DB1/c1	Ring upstairs DB1/c2	Immersion DB1/c3	lights down DB1/c4	Lights up DB1/c5	Ring down DB1/c6	Ring kitchen DB1/c8	Wall lights DB1/c9	Main bonding	Supplementary bond	
Number of points served			1	12	1	8	7	8	6	7	X	X	
Protective Device	Device BS		88	88	88	88	88	88	88	88	X	X	
	Type		2	2	2	2	2	2	2	2	X	X	
	Rating (A)		32	32	16	6	6	32	32	6	X	X	
Phase (colour or 3 phase)			R	R	R	R	R	R	R	R	X	X	
Cable size	Phase (mm²)		10	2.5	2.5	1.0	1.0	2.5	2.5	1.0	X	X	
	CPC (mm²)		4.0	1.5	1.5	1.0	1.0	1.5	1.5	1.0	X	X	
Continuity	Bonding (Ω)		X	X	X	X	X	X	X	X	0.1	0.2	
	CPC (Ω)		0.2	0.6	0.5	0.7	0.7	0.7	0.5	0.7	X	X	
	Radial	Design R1 + R2 (Ω)	0.4	X	0.9	1.0	1.0	X	X	1.0	X	X	
		Measured R1 + R2 (Ω)	0.3	X	0.7	0.9	0.8	X	X	0.8	X	X	
	Ring	Design R1 + R2 (Ω)	X	0.9	X	X	X	0.9	0.8	X	X	X	
		Measured R1 + R2 (Ω)	X	0.8	X	X	X	0.7	0.6	X	X	X	
Confirmation of ring Y or N			X	Y	X	X	X	Y	Y	X	X	X	

RECORD OF TEST RESULTS AND DESIGN DETAILS

Insulation resistance in MΩ		Phase to CPC	999	999	999	999	999	999	999	999	X	X	X
		Phase to neutral	999	999	999	999	999	999	999	999	X	X	X
		Neutral to CPC	999	999	999	999	999	999	999	999	X	X	X
		L1 to CPC	X	X	X	X	X	X	X	X	X	X	X
		L2 to CPC	X	X	X	X	X	X	X	X	X	X	X
		L3 to CPC	X	X	X	X	X	X	X	X	X	X	X
		L1 to neutral	X	X	X	X	X	X	X	X	X	X	X
		L2 to neutral	X	X	X	X	X	X	X	X	X	X	X
		L3 to neutral	X	X	X	X	X	X	X	X	X	X	X
Separation of circuits			999										
Polarity (Pass or Fail)			P	P	P	P	P	P	P	P	X	X	X
Earth fault loop Zs (Ω)		Maximum design Zs (Ω)	1.5	1.0	2.14	5.71	5.71	1.0	1.0	5.71	X	X	X
		Actual Measured Zs (Ω)	0.7	1.0	1.1	1.0	1.0	0.9	0.9	0.9	X	X	X
RCD		Rated IΔN (mA)	X	X	X	X	X	30	X	X	X	X	X
		½ ΔN Y/N	X	X	X	X	X	N	X	X	X	X	X
	Trip time (ms) at	IΔN trip ms 0°	X	X	X	X	X	70	X	X	X	X	X
		IΔN trip ms 180°	X	X	X	X	X	69	X	X	X	X	X
		150mA ms 0°	X	X	X	X	X	X	X	X	X	X	X
		150mA ms 180°	X	X	X	X	X	X	X	X	X	X	X
		Test button Y/N	X	X	X	X	X	Y	X	X	X	X	X
PEFC (Ia)		Origin	760	760	760	760	760	760	760	760	X	X	X
		Remote	328	230	209	230	230	255	255	255	X	X	X
PSCC (P-N)		Origin	910	910	910	910	910	910	910	910	X	X	X

ELECTRICAL INSTALLATION CERTIFICATE

DETAILS OF THE CLIENT

INSTALLATION ADDRESS

DESCRIPTION AND EXTENT OF THE INSTALLATION	New installation ☐
Description of installation:	Addition to an existing installation ☐
Extent of installation covered by this certificate:	Alteration to an existing installation ☐

FOR DESIGN, CONSTRUCTION & TESTING
I being the person responsible for the design, construction, inspection & testing of the elecrical installation (as indicated by my signature below). Particulars of which are described above, have exercised reasonable kill and care when carrying out the design, construction, inspection & testing hereby CERTIFY that the said work for which I have been responsible is to the best of my knowledge and belief in accordance with BS7671.............., ammended to ....(date)

Details and departures from BS7671 as ammended (regulations 120-02, 120-05)

The extent of liability of the signatory is limited to the work described as the subject of this certificate.
Name (IN BLOCK LETTERS):............. .... Position:..
Signature:.. Date:..
For and on behalf of:..
Address:..
....Postcode:.............................. Tel:..

NEXT INSPECTION
I recommend that this installation is further inspected and tested after an interval of not more than ....years/months

SUPPLY CHARACTERISTICS AND EARTHING ARRANGEMENTS

Earthing arrangements	**Number & type of live conductors**		**Nature of supply paramiters**	**Supply protective device characteristics**
TN- C ☐	a.c. ☐	d.c. ☐	Nominal voltage U/Uo..........V	
TN-S ☐	1-phase, 2 wire ☐	2-pole ☐	Nominal frequency, f.............Hz	Type:..............................
TN-C-S ☐	2-phase, 3 wire ☐	3-pole ☐	Prospective fault current, Ipf.......KA	Nominal current
TT ☐	3-phase, 3 wire ☐	Other ☐	External earth loop impedance, Ze........Ω	rating...........................A
IT ☐	3-phase, 4 wire ☐			

SUPPLY CHARACTERISTICS AND EARTHING ARRANGEMENTS

Means of earthing	**Maximum demand**
Suppliers facility ☐	Maximum demand (load)................................Amps
Installation earth electrode ☐	**Details of installation earth electrode** Type........................Location.......................Electrode resistance to earth......................Ω

Main protective conductors

Earthing conductor: Materiel.............................CSA.........................connection verified ☐
Main equipotential bonding conductors: Materiel.............................CSA.........................connection verified ☐

To incoming water and/or gas service ☐ **To other elements..**

Main switch or circuit-breaker

BS, Type..............No of poles.................Current rating..............A Voltage rating......................V
location..Fuse rating or setting..A
Rated residual operating current $I\Delta n$ = ...mA, and opperating time of ....ms (at $I\Delta n$)

COMMENTS ON EXISTING INSTALLATION

SCHEDULES:
The attached schedules are part of this document and this certificate is valid only when they are attached to it.
.........................Schedules of inspection and..............................Schedules of Test Results are attached.

RECORDING YOUR RESULTS

Something that must not be forgotten is that the record of results form a legal document. You are personally liable so make sure that you carry out the tests correctly and record the information accurately.

I have given you an example of a completed record of result on pages 64 and 65, and something on this line may be adequate for a domestic installation, but when it comes to commercial or industrial records you will need to complete a record of results for each and every distribution board tested.

It is also a legal requirement that you have your instruments calibrated so that the results collected are accurate.

Now if you look at pages 64 and 65 there is information that I have not yet covered, that is apart from entering your name and date of the tests. You may need to consult the local electricity board for the external impedance (at point D) and possibly the earthing arrangements for the installation (at point C), although the earthing arrangements can be sought by looking at the incoming supply and the external impedance Ze can be measured as in the test on page 46.

You will also have to enter the make and serial numbers for the instruments you are using (at point F and G).

In some of the rows you will see the words DESIGN (at points R, T, and HH). The figures that you enter here should be available from the original design specification.

Now on page 68 are the declarations that you will have to sign stating that you are the person responsible for the testing and that all of the documentation is present and correct and if there are any recommendations such as remedial repair work to specific items or a complete re-wire.

Please note that the forms I have suggested are in no way complete and are only samples that should be used as a basic guide for your results.

CARRYING OUT ALTERNATIVE TESTS FOR SEPARATION OF CIRCUITS

If you remember back on page 36, I was discussing how long it would take to carry out tests to confirm separation of circuits on a 6 way board. Well lets look at the practicalities;

In the consumer unit there are 6 circuits each having 3 conductors phase, neutral and cpc. Therefore there are 18 conductors to check, the meter has two test probes so effectively we are looking for the number of combinations that you can select 2 from 18. The formula goes something like this,

$$ {}^{n}C_{r} = \frac{n}{r!\,(n-r)!} $$

Now we can substitute our figures (remember 2 from 18)

$$ {}^{18}C_{2} = \frac{18 \times 17}{2} = 153 $$

So there you have it, you are going to have to make 153 separate tests. How long are you going to allow for the tests? Are you going to label all of the conductors? How are you going to guarantee that you test all of the conductors?

One solution is to couple all of the conductors for each circuit together, so now you have reduced the number of tests to 15. That is you are now testing 2 from 6 because you have strapped the phase, neutral and cpc of each circuit together. You can do this because previously you carried out insulation resistance of the individual circuit (test 3).

So if you measured the insulation resistance between phase, neutral and cpc for the individual circuits, you can now strap them together (see Fig: 30 page 71).

Test: Alternative measurement for separation of circuits
Reason: To check for damage to cables and a breakdown in insulation and interconnections between circuits.

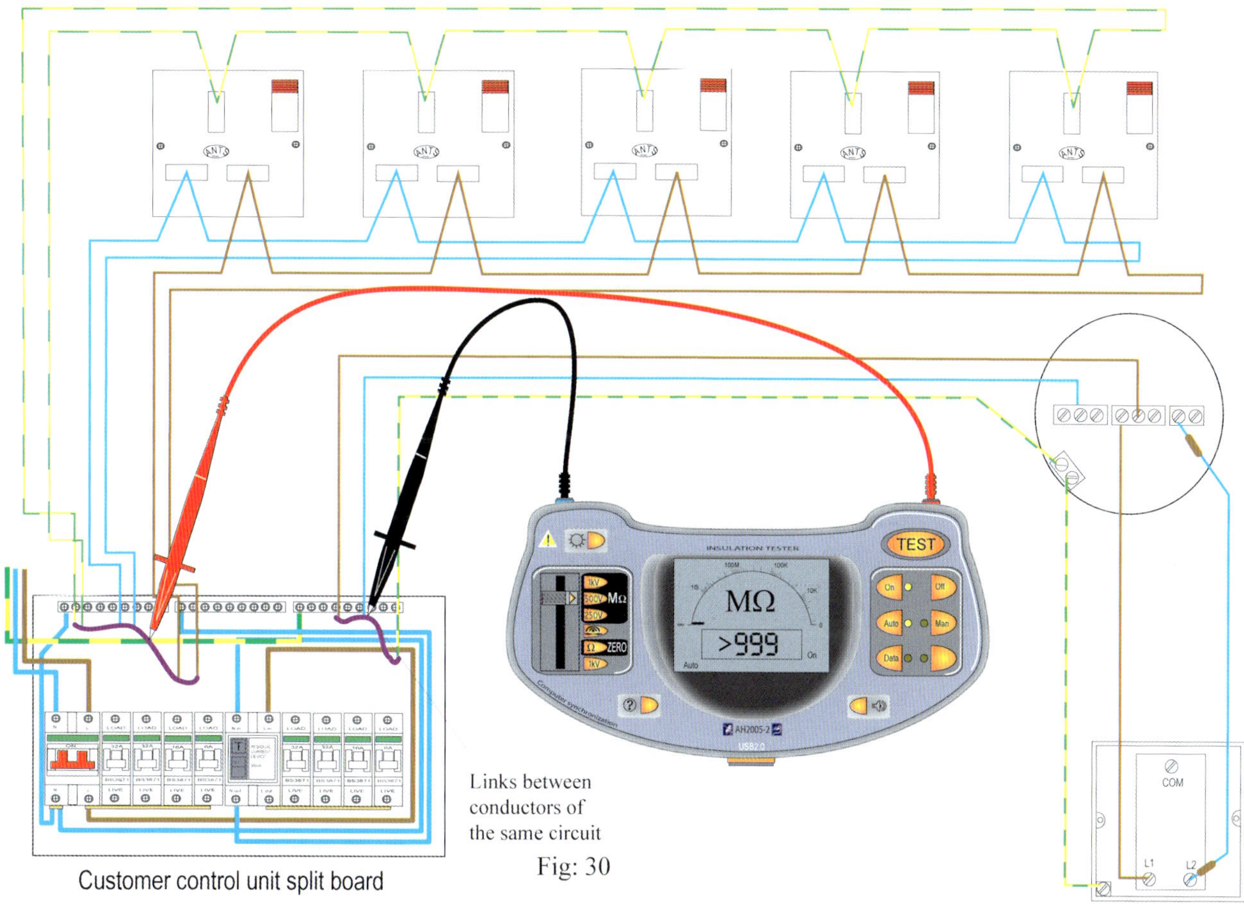

Fig: 30

UNDERSTANDING HOW Zs IS FORMED IN INDUSTRIAL INSTALLATIONS

Depicted in Fig: 31 on page 73 you will see an example of the typical arrangement of a final circuit in an industrial application.

We know now that the formula for earth loop impedance is:

$$Zs = Ze + (R1 + R2)$$

So where do these fit in with the measurement and how do we know whether the reading you measured is right? Depicted in Fig: 31 you will see the letter A depicting the starting point of the circuit that ends at the motor. The sub-main feeding distribution board B has to figure in the calculations because of the added resistance.

So if you are going to measure Zs for the motor circuit you would carry out the test as described in Fig: 20 page 45, but you are in fact measuring the external impedance Ze, the R1 + R2 value of the sub-main from point A to point B and the R1 + R2 value from point B to point C, so the formula for circuits incorporating sub-mains is.....

$$Zs = Ze + (R1 + R2) + (R1 + R2)$$

It states in the regulations that earth loop should be measured at the origin and at the remote points in the circuit. The remote point in this case it is at the motor, but the origin of the circuit is at the distribution board B. So what about the sub-main? Well this is a separate circuit on its own and all of the necessary tests should be carried out for this circuit.

Example

Lets assume you measured the Zs for the motor circuit in Fig: 31 page 73 and arrived at a Zs = 0.96Ωs. You know that the circuit should disconnect in 5 seconds and assuming the protective device is a 10A BS 3871 type 2, the regulations state the maximum value for Zs is 3.43Ωs. As the figure that you measured is less than 3.43Ωs the circuit passes. You have actually measured the external impedance Ze, the R1+R2 value for the sub-main from A-B and the R1+R2 value for the final circuit from point B-C. If you measure the Zs at point B this in turn can be used as the external impedance for distribution board B.

When the designer of the installation receives the test report they can check to see if the installation has been installed as it was designed. If the results don't match the design, then the designer knows that it has not been installed as it was intended and will investigate to find out why.

Understanding how Zs is made up in industrial installations

Origin and remote points refers to where the cable terminates

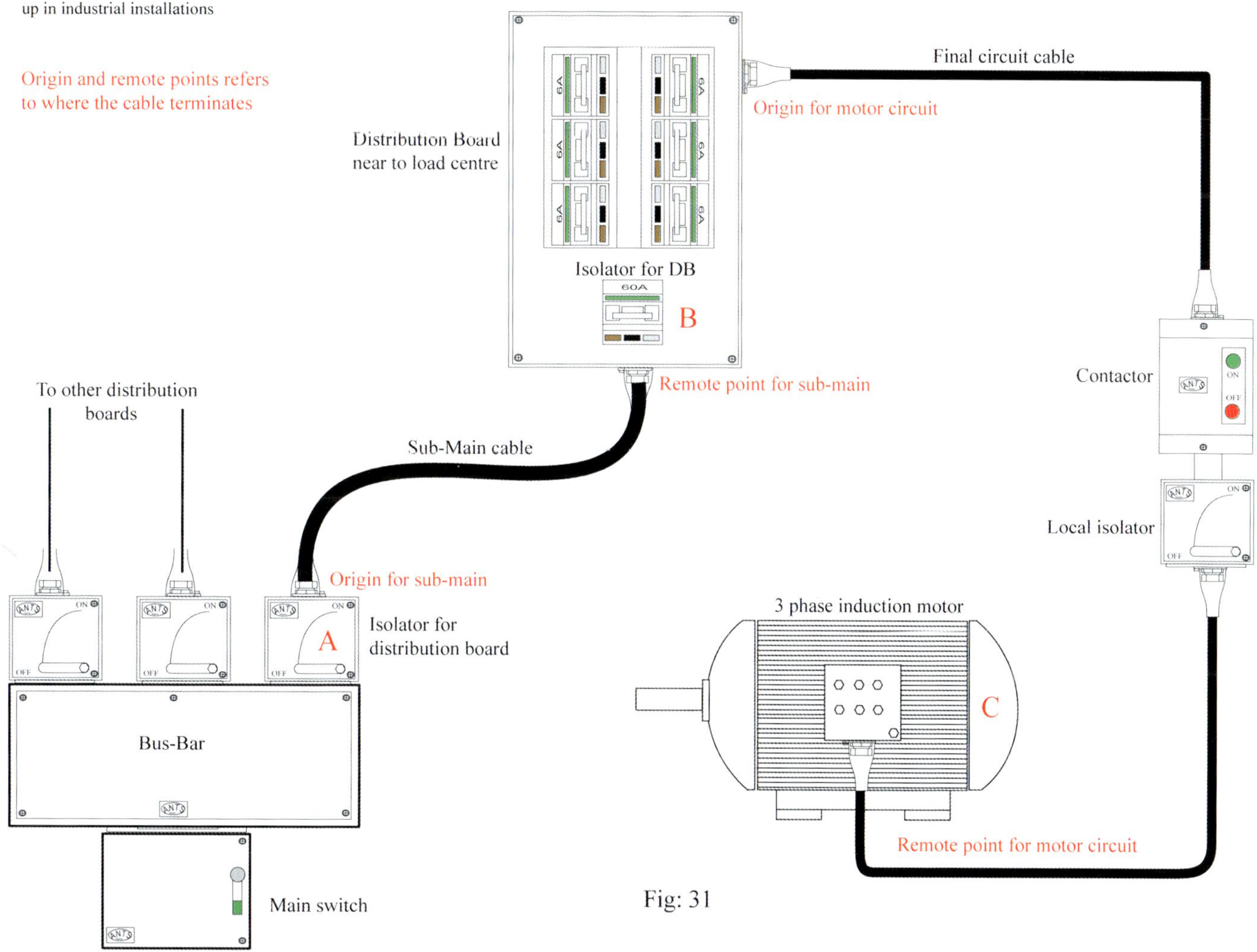

Fig: 31

TESTING INSULATION RESISTANCE AT THE MAIN TAILS

Depicted in Fig: 32 on page 75 is an example of checking Insulation resistance at the main tails of the distribution board.

The sequence is as follows:

- Remove all loads
- Ensure Isolation
- Place links in circuit as indicated in Fig: 32
- Select an Insulation continuity tester and set to measure 500V
- Place instrument on the circuit as shown
- Take reading and record

The reason for carrying out insulation resistance test at the main tails is to save time. If all of the circuits are connected and turned on then you will test all of the circuits at the same time.

When circuits are tested in parallel, the reciprocal of the insulation resistances in some instances, can appear to make the circuits fail the test. If this is the case you will then need to split the circuits up and test them individually as in Fig: 13, 14 and 15.

This one test can save a considerable amount of time especially with commercial and industrial distribution boards.

Warning: You will only be able to carry out the tests in this way if the tails can be isolated.

Remember to remove all links before reinstating the supply

This table gives the minimum values allowed

CIRCUIT NOMINAL VOLTAGE (V)	TEST VOLTAGE d.c. (V)	MINIMUM INSULATION RESISTANCE (MΩ)
SELV AND PELV *	250	0.25
UP TO AND INCLUDING 500V WITH THE EXCEPTION OF THE ABOVE SUPPLIES	500	0.5
ABOVE 500V	1000	1.0

*SELV = Separated Extra-low Voltage PELV = Protected Extra-low Voltage

Test: Insulation resistance at the main tails

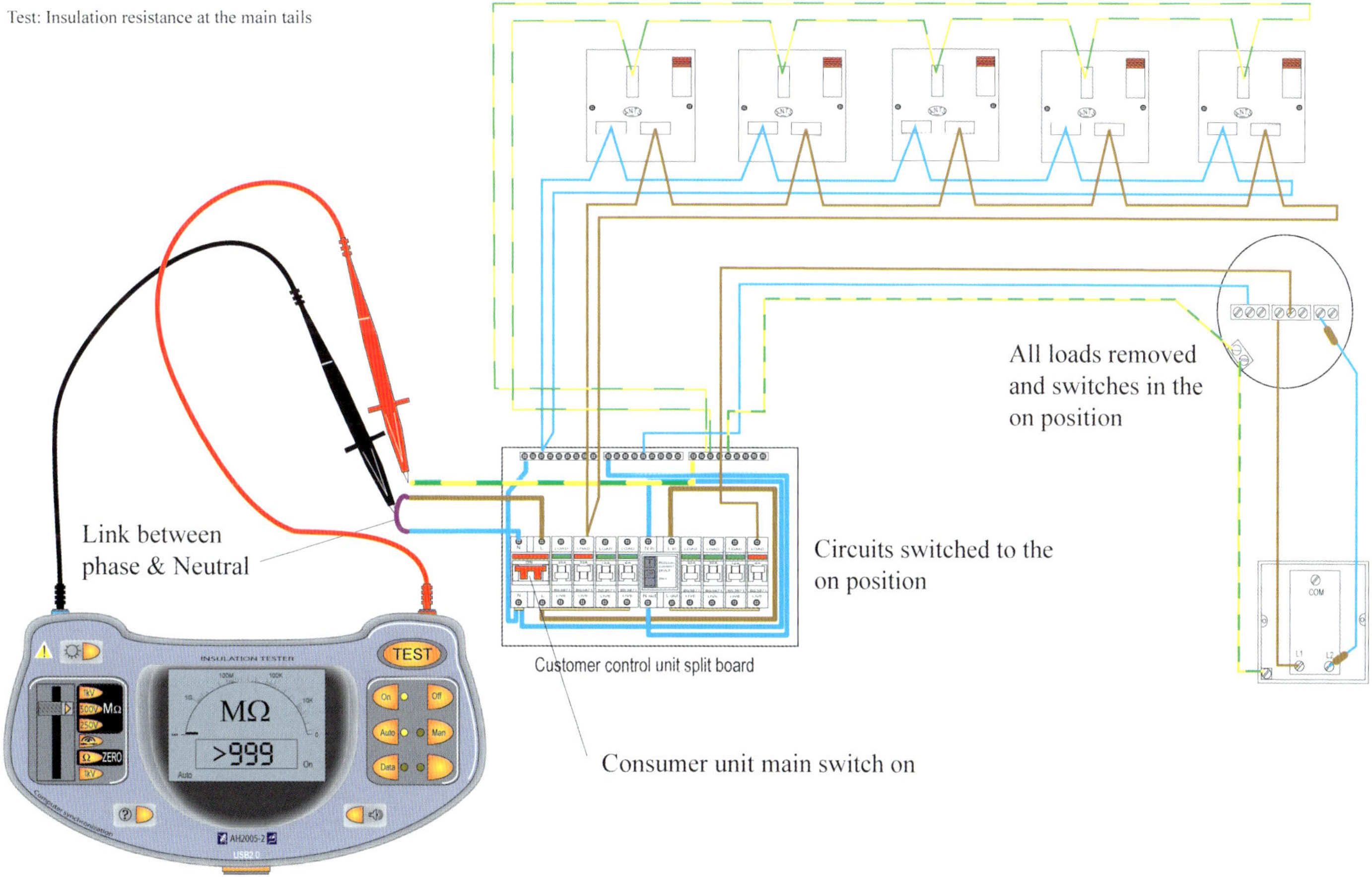

Fig: 32

INSULATION TEST ON SELV AND PELV

Depicted in Fig: 33 on page 77 is an example of checking Insulation resistance on a separated extra low voltage source. The same test can be carried out on protected extra low voltage sources as well.

All loads must be disconnected and switches turned on. Remember to remove any electronic devices in the circuit.

The sequence is as follows:

- Remove all loads
- Ensure Isolation
- Select an Insulation continuity tester and set to measure 500V (see note 1)
  This test must be sustained for 60 seconds
- Place instrument on the circuit as shown
- Take reading and record

The reason for carrying out insulation resistance test is to see if the insulation of the SELV source is separated between the low voltage side and the extra-low voltage side. If you apply a voltage with the test instrument you are trying to encourage electron flow across the insulation of the conductors, if electrons flow then there is a breakdown in the insulation and the resistance can be measured.

Note: 1 I have opted to use a secondary voltage on the SELV source of 50V rms. so therefore you will need to use a test voltage of 500V as stated in Regulation 411-02-09, however if the secondary voltage was 25V then you would test at 250V.

This table gives the minimum values allowed

CIRCUIT NOMINAL VOLTAGE (V)	TEST VOLTAGE d.c. (V)	MINIMUM INSULATION RESISTANCE (MΩ)
SELV AND PELV * (up to 25V rms or 60V ripple-free dc)	250	0.25
UP TO AND INCLUDING 500V WITH THE EXCEPTION OF THE ABOVE SUPPLIES	500	0.5
ABOVE 500V	1000	1.0

*SELV = Separated Extra-low Voltage PELV = Protected Extra-low Voltage

There are additional tests to be carried out so you will need to consult the IEE Guidance notes on testing.

Test: Insulation resistance between primary and secondary windings of an SELV source

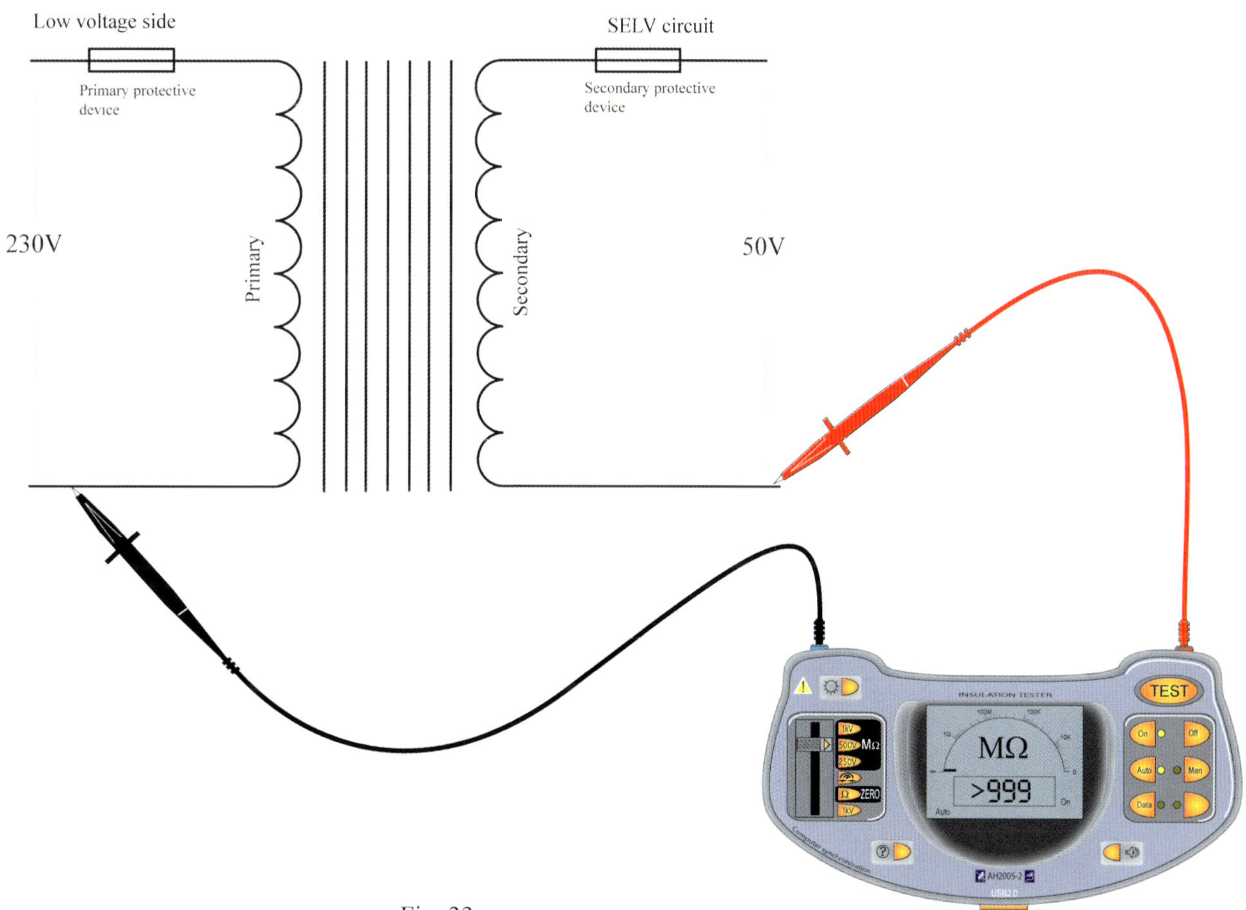

Fig: 33

GENERAL NOTES OF IMPORTANCE

I hope that this book has clarified the way in which you test installations and that it will become an invaluable guide.

1. All references to HSE GS38 are purely the reference to the specification that test equipment must conform to. The lead sets referred to in the text and drawings are special leads available from the instrument manufacturers.

2. **Remember to remove all links that have been used before reinstating the supply**

3. Conductor sequence refers to the sequence that the conductors have been terminated in the distribution board. You should be aware that the largest load is terminated next to the main switch, this is also known as circuit 1, so the corresponding neutral and cpc should be terminated in position 1 on the neutral and earthing bar in the distribution board. This will help locate all the conductors for circuits when testing and altering the installation.

4. Alterations to the installation must receive the same tests that a complete, new or old installation receives. For arguments sake lets assume that you have been asked to install a fused spur to supply a dishwasher. An assessment of the supply and circuit conductors must be made to allow for the extra current demand. When the work has been undertaken the whole circuit and not just the alteration must be subjected to all of the tests laid down in the Regulations, as you will assume responsibility for the circuit that you have altered.

5. I have referred to voltages as 400V and 230V these are to come in line with the amendments to the Regulations (AMD 8536).

6. In test 7 Functional testing pages 48 and 50 the 150mA tripping current test only needs to be carried out if the RCD is providing supplementary protection from direct contact.

7. Direct contact means the touching of a live part, and indirect contact means the touching of a conductive part made live through fault conditions.

8. PSCC is the measurement of the Prospective Short Circuit Current and PEFC is the measurement of the Prospective Earth Fault Current. Any instrument that can measure either of these is capable of both tests, if not calculations will need to be carried out as explained in the text.